Contents

STUDY LINK 1·1 Unit 1: Family Letter

Introduction to *Sixth Grade Everyday Mathematics*®

The program we are using this year—*Everyday Mathematics*—offers students a broad background in mathematics. Some approaches in this program may differ from the ones you learned as a student. That's because we're using the latest research results and field-test experiences to teach students the math skills they'll need in the 21st century. Following are some program highlights:

◆ A problem-solving approach that uses mathematics in everyday situations

◆ Activities to develop confidence, self-reliance, and cooperation

◆ Repeated review of concepts throughout the school year to promote mastery

◆ Development of concepts and skills through hands-on activities

◆ Opportunities to communicate mathematically

◆ Frequent practice using games as alternatives to tedious drills

◆ Opportunities for home and school communication

Sixth Grade Everyday Mathematics emphasizes a variety of content.

Number Relations

◆ Recognizing place value in whole numbers and decimals

◆ Using exponential and scientific notation

◆ Finding factors and multiples

◆ Converting between fractions, decimals, and percents

◆ Ordering positive and negative numbers

Operations, Computation, and Mental Arithmetic

◆ Solving problems involving whole numbers, fractions, decimals, and positive and negative numbers

◆ Applying properties of addition, subtraction, multiplication, and division

Data and Chance

◆ Collecting, organizing, displaying, and analyzing data

◆ Identifying and comparing landmarks of data sets (mean, median, mode, and range)

◆ Using probability to represent and predict outcomes and analyze chance

Measurement, Measures, and Numbers in Reference Frames

◆ Measuring using metric and U.S. customary units

◆ Using formulas to calculate circumference, area, and volume

◆ Naming and plotting points on a coordinate grid

Geometry

◆ Measuring and drawing angles

◆ Understanding properties of angles

◆ Identifying and modeling similar and congruent figures

◆ Constructing figures with a compass and a straightedge

◆ Drawing to scale

◆ Exploring transformations of geometric shapes

◆ Experimenting with modern geometric ideas

Patterns, Functions, and Algebra

◆ Creating and extending numerical patterns

◆ Representing and analyzing functions

◆ Manipulating algebraic expressions

◆ Solving equations and inequalities

◆ Working with Venn diagrams

◆ Applying algebraic properties

◆ Working with ratios and proportions

Throughout the year, you will receive Family Letters telling you about each unit. Letters may include definitions and suggestions for at-home activities. Parents and guardians are encouraged to share ideas pertaining to these math concepts with their child in their home language. You and your child will experience an exciting year filled with discovery.

Building Skills Through Games

Games are as integral to the *Everyday Mathematics* program as Math Boxes and Study Links because they are an effective and interactive way to practice skills.

In this unit, your child will work on understanding place value of whole and decimal numbers, data landmarks, and order of operations by playing the following games.

Detailed game instructions for all sixth-grade games are provided in the *Student Reference Book.*

High-Number Toss (Whole Number and Decimal Versions) See *Student Reference Book,* pages 323 and 324.

Students practice reading and comparing whole numbers through hundred-millions and decimals through thousandths.

Landmark Shark See *Student Reference Book,* pages 325 and 326.

Students practice finding the mean, median, mode(s), and range of a set of numbers.

Name That Number See *Student Reference Book,* page 329.

Students practice writing number sentences using order of operations.

Collection, Display, and Interpretation of Data

Everyday Mathematics will help your child use mathematics effectively in daily life. For example, the media—especially newspapers and magazines—use data. Employees and employers need to know how to gather, analyze, and display data to work efficiently. Consumers need to know how to interpret and question data presented to them so they can make informed choices. Citizens need to understand government data to participate in the running of their country.

In *Everyday Mathematics,* data provide a context for the development of numeric skills that, in traditional programs, would be developed in isolation. In Unit 1, your child will work with data displayed in stem-and-leaf plots, circle graphs, step graphs, broken-line graphs, bar graphs, and tables.

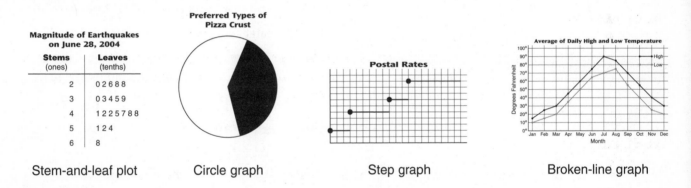

| Stem-and-leaf plot | Circle graph | Step graph | Broken-line graph |

The displays above relate to earthquake magnitudes, preferred pizza crusts, postal rates, and temperatures. Real-world applications support and enrich other areas of mathematics as well.

Throughout Unit 1, your child will look for graphs and tables in newspapers and magazines and bring them to school with your permission. The class will think critically about the materials collected. Students will consider the following questions:

◆ What is the purpose of the graph or table?

◆ Is the display clear and attractive, or can it be improved?

◆ Does the display seem accurate, or is it biased?

◆ Can you draw conclusions or make predictions based on the information?

Finally, students will learn a new game, *Landmark Shark,* which will help them develop skill in finding landmarks of data in various data sets. Ask your child to teach you how to play this game.

This should be a stimulating year, and we invite you to share the excitement with us!

Please keep this Family Letter for reference as your child works through Unit 1.

As You Help Your Child with Homework

As your child brings assignments home, you might want to go over the instructions together, clarifying them as necessary. Some of the answers listed below will guide you through the unit's Study Links.

Study Link 1·2

2. 90

5. Sample answers: Title: Weekly Allowance;
Unit: Dollars

6. 90   **7.** 80   **8.** 120   **9.** 80

Study Link 1·3

2. a. 4.8   **b.** 2.8, 4.2, 4.8   **c.** 4.1

3. 80   **4.** 110   **5.** 500   **6.** 50

Study Link 1·4

1. Mia: 80; Nico: 80   **3.** Mia: 80; Nico: 75

4. Mia: 25; Nico: 45

6. $5.82   **7.** $30.27

8. $14.24   **9.** $20.50

Study Link 1·5

1. a. 38   **b.** 147.5   **c.** 149.2

2. a. 29   **b.** 149   **c.** 151.3

3. $9.01   **4.** $1,107.47

5. $45.87   **6.** $35.67

Study Link 1·6

2. 90°F   **3.** About 25 minutes

4. Sample answers: **a.** About 100 minutes

b. The rate of cooling levels off to $2\frac{1}{2}$°F every 10 min.

5. a. no

b. The tea cools very quickly at first, but then the temperature drops slowly.

6. 1,728   **7.** 3,306   **8.** 4,484   **9.** 2,538

Study Link 1·7

2. 5   **3.** 2   **4.** 3 times   **5.** 2 times

6. 2; 3   **8.** 6,613   **9.** 8,448   **10.** 10,872

11. 9, 711

Study Link 1·8

2. $1.29   **3. a.** $1.75

b. Sample answer: The price difference per ounce is $0.23. The price jumps another $0.23 for every additional part of an ounce.

5. 28   **6.** 45   **7.** 67   **8.** 55

Study Link 1·9

1. Answers vary.

3. men   **4. a.** 89%   **b.** 11%

5. 10% greater   **6.** 60% greater

7. Sample answer: Because they don't know the person, they don't know how the stranger will react.

8. 24   **9.** 14   **10.** 32   **11.** 19

Study Link 1·10

1. Width (ft): 2; 3; 4; 6; 8; 9
Area (ft²): 20; 27; 32; 36; 32; 27; 11

2. square

3. Length (yd): 24; 16; 12; 8; 6; 4; 2; 1
Perimeter (yd): 98; 52; 38; 32; 28; 32; 38; 98

4. a. 6 yd or 8 yd   **b.** 8 yd or 6 yd   **5.** $0.10

6. $4.00   **7.** $485.00   **8.** $2,050.00

Study Link 1·11

1. 165,000   **2.** 2003 and 2004

3. Sample answer: Yes. The population in 2005 would have to be 310,000 for the claim to be true.

4. $5.00   **5.** $90.00   **6.** $13,925.00   **7.** $0.89

Study Link 1·12

1. a. 30 min

b. 1 hr 20 min, or $1\frac{1}{3}$ hours, or 80 min

2. 2 hr 20 min, or $2\frac{1}{3}$ hours, or 140 min

3. Sample answer: Biased. There are other ways to get to work, so not all commuters are represented.

4. $70.00   **5.** $8.45   **6.** $25.92

4

STUDY LINK 1·2 | **Mystery Line Plots and Landmarks**

1. Draw a line plot for the following spelling test scores.

   100, 100, 95, 90, 92, 93, 96, 90, 94, 90, 97

2. The mode of the above data is _____.

3. Draw a line plot below that represents data with the following landmarks.
   Use at least 10 numbers.

   range: 7   minimum: 6   median: 10   modes: 8 and 11

4. Describe a situation in which the data in the above line plot might occur.

   _____

   _____

5. Give the line plot a title and a unit.

   Title _____

   Unit _____

Practice

6. 540 ÷ 6 = _____

7. 7,200 ÷ 90 = _____

8. 84,000 ÷ 700 = _____

9. 400,000 ÷ 5,000 = _____

STUDY LINK 1·3 Stem-and-Leaf Plots

Every day, there are many earthquakes worldwide. Most are too small for people to notice. Scientists refer to the size of an earthquake as its magnitude. Earthquakes are classified in categories from minor to great, depending on magnitude.

Class	Magnitude
Great	8.0 or more
Major	7–7.9
Strong	6–6.9
Moderate	5–5.9
Light	4–4.9
Minor	3–3.9

The table below shows the magnitude of 21 earthquakes that occurred on June 28, 2004.

Magnitude of Earthquakes Occurring June 28, 2004										
4.2	5.2	2.8	4.8	3.9	2.0	3.3	4.8	4.5	3.5	2.2
2.6	3.4	6.8	3.0	4.7	2.8	4.2	4.1	5.4	5.1	

1. Construct a stem-and-leaf plot of the earthquake magnitude data.

2. Use your stem-and-leaf plot to find the following landmarks.

   a. range _____

   b. mode(s) _____

   c. median _____

Magnitude of Earthquakes Occurring on June 28, 2004

Stems (ones)	Leaves (tenths)

Practice

3. 6,400 ÷ 80 = _____

4. 121,000 ÷ 1,100 = _____

5. 3,000,000 ÷ 6,000 = _____

6. 600,000 ÷ 12,000 = _____

STUDY LINK 1·4 | Median and Mean

Mia's quiz scores are 75, 70, 75, 85, 75, 85, 80, 95, and 80.

Nico's quiz scores are 55, 85, 95, 100, 75, 75, 65, 95, and 75.

1. Find each student's mean score. Mia _____ Nico _____

2. Make a stem-and-leaf plot for each student's scores.

a. Mia's Quiz Scores

Stems (100s and 10s)	Leaves (1s)

b. Nico's Quiz Scores

Stems (100s and 10s)	Leaves (1s)

3. Find each student's median score. Mia _____ Nico _____

4. What is the range of scores for each student? Mia _____ Nico _____

5. Which landmark, mean or median, is the better indicator of each student's overall performance? Explain.

_____

_____

_____

_____

_____

Practice

6. $4.57 + $1.25 = _____

7. $14.49 + $15.78 = _____

8. $19.99 − $5.75 = _____

9. $39.25 − $18.75 = _____

9

STUDY LINK 1·5 | Range, Median, and Mean

Heights were measured to the nearest centimeter for
12 boys and 12 girls. All of the students were 12 years old.

Boys' heights: 157, 150, 131, 143, 147, 169, 148, 147, 145, 163, 139, 151

Girls' heights: 146, 164, 138, 149, 145, 167, 150, 156, 143, 148, 149, 160

1. Make a stem-and-leaf plot for the boys' data.
   Then find the range, median, and mean of the
   boys' heights.

   a. range _____

   b. median _____

   c. mean   _____

   Boys' Heights (cm)

Stems (10s)	Leaves (1s)

2. Make a stem-and-leaf plot for the girls' data.
   Then find the range, median, and mean of the
   girls' heights.

   a. range _____

   b. median _____

   c. mean   _____

   Girls' Heights (cm)

Stems (10s)	Leaves (1s)

Practice

3. $5.86 + $3.15 = _____

4. $221.17 + $886.30 = _____

5. $75.37 − 29.50 = _____

6. $124.35 − $88.68 = _____

STUDY LINK 1·6 | Cooling Off

The graph shows how a cup of hot tea cools as time passes.

1. Use the graph to fill in the missing data in the table.

2. What is the tea's approximate temperature after 30 minutes? _____

3. About how many minutes does it take for the tea to cool to a temperature of 95°F?

_____

4. a. About how many minutes do you think it will take the tea to cool to room temperature (70°F)?

_____

_____

b. Why do you think so?

_____

_____

_____

5. a. Does the tea cool at a constant rate? _____

b. Explain your answer.

_____

_____

_____

Elapsed Time (minutes)	Temperature (°F)
0 (pour tea)	
10	
40	
	100
	115
5	

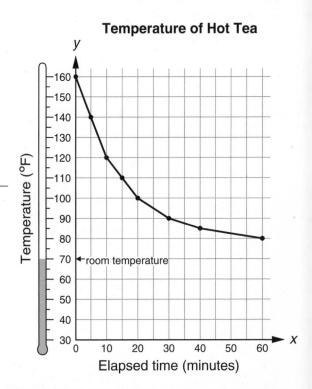

Temperature of Hot Tea

Practice

6. 32 * 54 = _____

7. _____ = 87 * 38

8. 59 * 76 = _____

9. _____ = 94 * 27

13

STUDY LINK 1·7 | Using Bar Graphs

Every week, Ms. Penczar gives a math quiz to her class of 15 students. The table below shows the class's average scores for a six-week period.

1. Draw a bar graph that shows the same information. Give the graph a title and label each axis.

   Use the graph you drew to answer the following questions.

Week	Class Average
1	68
2	66
3	79
4	89
5	91
6	88

2. The highest average score occurred in

   Week _____.

3. The lowest average score occurred in

   Week _____.

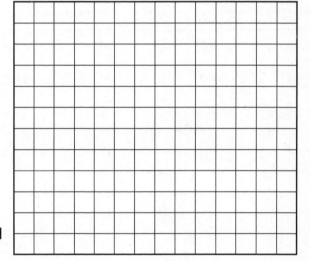

4. How many times did scores improve from one week to the next?

   _____

5. How many times did scores decline from one week to the next?

   _____

6. The greatest one-week improvement occurred

   between Week _____ and Week _____.

7. Name a possible set of scores for Ms. Penczar's 15 students that would result in the class average given for Week 2.

   _____ _____ _____ _____ _____ _____ _____ _____

   _____ _____ _____ _____ _____ _____ _____

Practice

8. 389 * 17 = _____

9. _____ = 176 * 48

10. 453 * 24 = _____

11. _____ = 249 * 39

**STUDY LINK
1·8** **The Cost of Mailing a Letter**

The cost of mailing a first-class letter in the United States depends on how much the letter weighs. The table at the right shows first-class postal rates in 2004: 37 cents for a letter weighing 1 ounce or less; 60 cents for a letter weighing more than 1 ounce but not more than 2 ounces; and so on.

2004 First-Class Postal Rates	
Weight (oz)	**Cost**
1	$0.37
2	$0.60
3	$0.83
4	$1.06
5	$1.29
6	$1.52

A step graph for these data has been started on page 23. Notice the placement of dots in the graph. For example, on the step representing 60 cents, the dot at the right end, above the 2, shows that it costs 60 cents to mail a letter weighing exactly 2 ounces. There is no dot at the left end of the step—that is, at the intersection of 1 ounce and 60 cents—because the cost of mailing a 1-ounce letter is 37 cents, not 60 cents.

1. Continue the graph for letters weighing up to 6 ounces.

2. Using the rates shown in the table, how much would it cost to send a letter that weighs $4\frac{1}{2}$ ounces? _____

Try This

3. a. Using the rates shown in the table, how much would it cost to mail a letter that weighs $6\frac{1}{2}$ ounces? _____

b. How did you determine your answer?

_____

_____

_____

4. Continue the graph on page 23 to show the cost of mailing a first-class letter weighing more than 6 ounces, but not more than 7 ounces.

Practice

5. $\frac{252}{9}$ = _____

6. _____ = $8)\overline{360}$

7. $\frac{469}{7}$ = _____

8. _____ = $9)\overline{495}$

17

The Cost of Mailing a Letter *continued*

Cost of Mailing a First-Class Letter in the United States in 2004

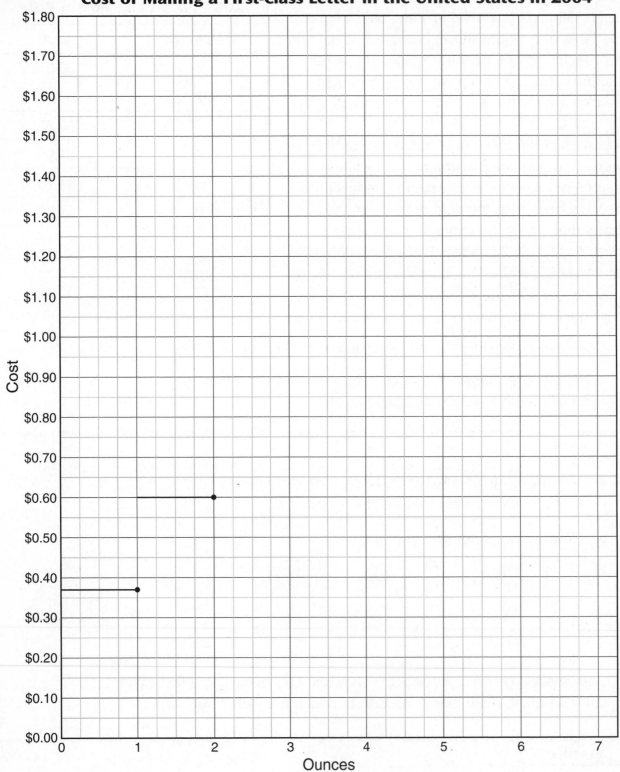

Name Date Time

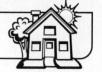

STUDY LINK 1•9

Analyzing Circle Graphs

SRB 145

1. Would you be willing to tell strangers that they had

smudges on their faces? yes no

food stuck between their teeth? yes no

dandruff? yes no

A marketing research company asked men and women these same questions. The results are summarized in the circle graphs below.

Use the legend to read the graphs. ▒ yes, would tell ☐ no, would not tell

2. Write estimates for the percents represented by each graph.

Smudge on Face

Women

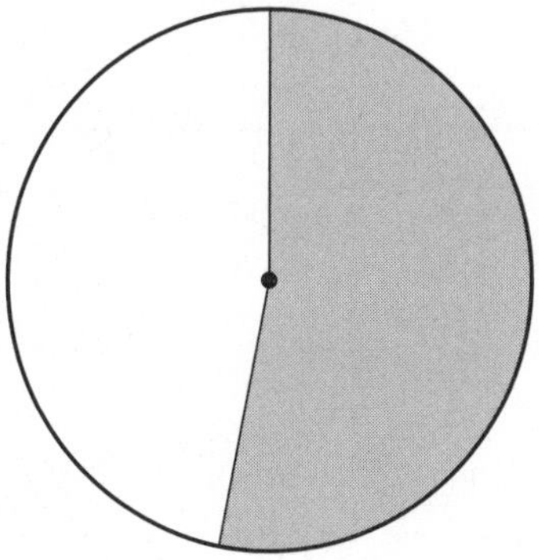

Estimates:

yes ________ no ________

Men

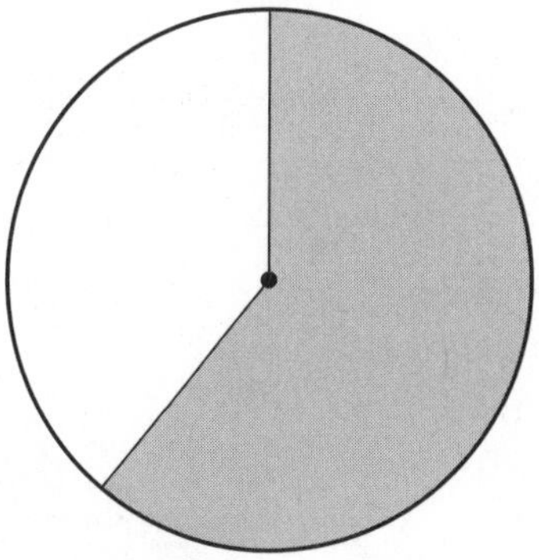

Estimates:

yes ________ no ________

Food in Teeth

Women

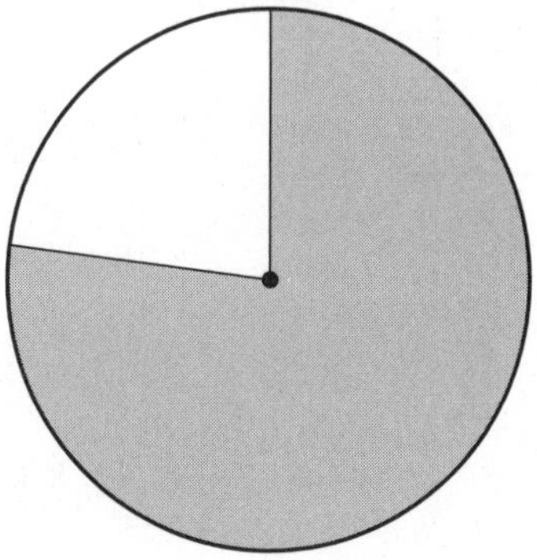

Estimates:

yes ________ no ________

Men

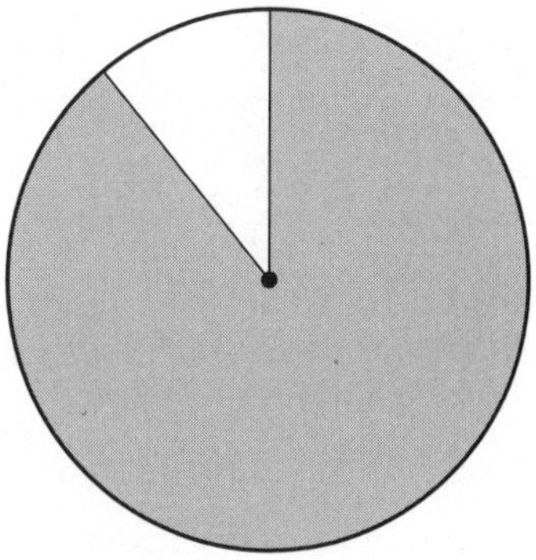

Estimates:

yes ________ no ________

Dandruff

Women

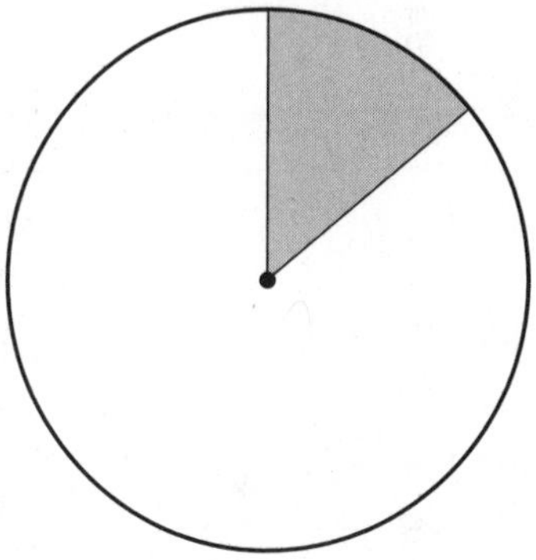

Estimates:

yes ________ no ________

Men

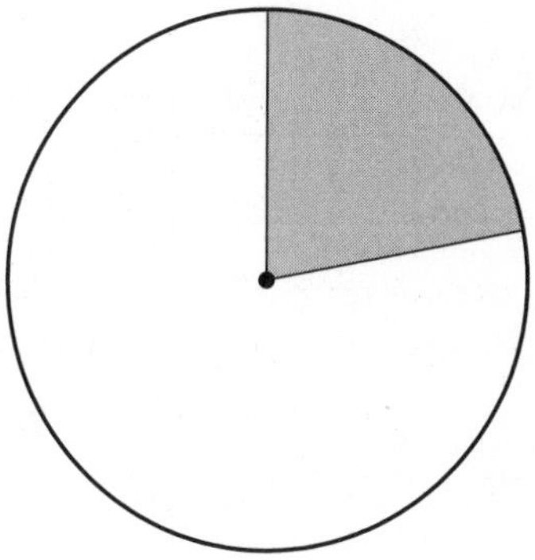

Estimates:

yes ________ no ________

Source: America by the Numbers

STUDY LINK 1·9 **Analyzing Circle Graphs** *continued*

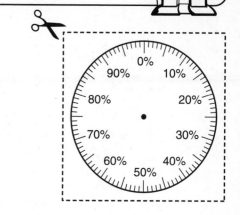

Cut out the Percent Circle at the right and poke a hole in the center with a pencil. Use the Percent Circle to find the percent represented by each sector mentioned in the questions below.

3. According to the survey, are men or women more likely to alert strangers to an embarrassing situation? _____

4. a. About what percent of men say they would tell strangers that they had food stuck between their teeth? _____

b. About what percent of men would not be willing to tell? _____

5. In the survey, how much greater is the percent of men who would be willing to alert strangers to smudges on their faces than the percent of women who would be willing to do so? _____

6. How much greater is the percent of women who would be willing to tell strangers about food in their teeth than the percent of women who would tell strangers about dandruff? _____

7. Why do you think people might be hesitant to alert strangers to such situations?

_____

_____

_____

Practice

8. $\frac{336}{14}$ = _____

9. _____ = $50\overline{)700}$

10. $\frac{992}{31}$ = _____

11. _____ = $29\overline{)551}$

21

STUDY LINK 1·10 Perimeter and Area

The student council is preparing the gym floor for the annual talent show. They will use 24 feet of tape to mark the seating area for the judges.

The table below lists the lengths of some rectangles with perimeters of 24 feet. Complete the table. You may want to draw the rectangles on grid paper. Let the side of each grid square represent 1 foot.

1.

Length (ft)	10	9	8	7	6	5	4	3	2	1
Width (ft)				5		7			10	11
Perimeter (ft)	24	24	24	24	24	24	24	24	24	24
Area (ft²)				35		35			20	

2. How would you describe the rectangular region that will provide the largest seating area for the judges? _____

The stage area for the talent show will be 48 square yards. The table below lists the lengths of some rectangles with areas of 48 yd². Complete the table.

3.

Length (yd)	48							3		
Width (yd)	1	2	3	4	6	8	12	16	24	48
Perimeter (yd)						28			52	
Area (yd²)	48	48	48	48	48	48	48	48	48	48

4. What is the length and width of the rectangular region that will take the least amount of tape to mark off?

a. length _____   **b.** width _____

Practice

5. $0.01 * 10 = _____

6. $0.40 * 10 = _____

7. $48.50 * 10 = _____

8. $205.00 * 10 = _____

STUDY LINK 1·11 | **The Population of River City**

The way a graph is constructed affects how fairly the data are represented.

The mayor of River City is trying to convince the city council that the city needs more schools. She claims that the city's population has doubled since 1998. The mayor used the graph below to support her claim.

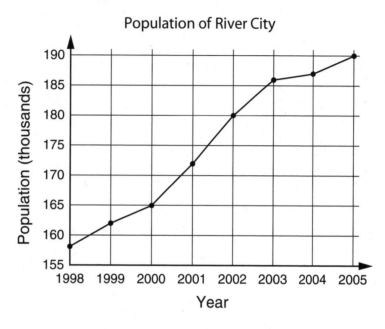

Population of River City

1. According to the graph, what was the population in 2000? _____

2. Between which 2 years was the increase in population the least?

   _____

3. Is the mayor's claim misleading? Explain.

   _____

   _____

   _____

Practice

4. $0.05 ∗ 100 = _____

5. $0.90 ∗ 100 = _____

6. $139.25 ∗ 100 = _____

7. _____ ∗ 100 = $89.00

Survey Results

A sample of 2,000 working adults was surveyed to determine how much time they spend performing certain weekly activities and how much time they would prefer to spend on these activities. For example, the average time adults spend on household chores is 4 hr 50 min, while the average time they prefer to do household chores is 2 hr 30 min.

The survey results are shown in the side-by-side bar graph below.

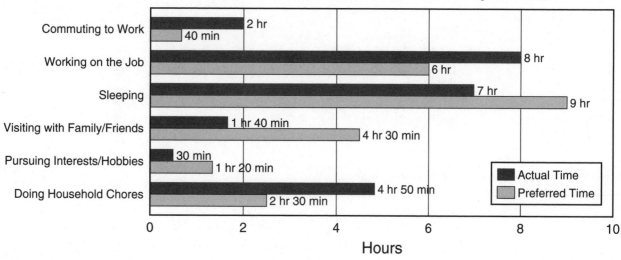

1. a. What is the actual time spent pursuing interests/hobbies? _____

   b. How much time would adults prefer to spend pursuing interests/hobbies?

   _____

2. What is the difference between the actual time and preferred time for doing household chores?

   _____

3. For this survey, researchers interviewed every 25th rider who boarded a commuter train on a Monday morning. Do you think this sampling method provides a random or biased sample? Explain.

   _____

   _____

| **Practice** |

4. $700.00 ÷ 10 = _____   **5.** $84.50 ÷ 10 = _____   **6.** $259.20 ÷ 10 = _____

STUDY LINK 1·13

Unit 2: Family Letter

Operations with Whole Numbers and Decimals

In Unit 2, your child will revisit operations with whole numbers and decimals from earlier grades and will continue strengthening previously developed number skills. We will work with estimation strategies, mental methods, paper-and-pencil algorithms, and calculator procedures with whole numbers. We will also develop techniques for working with decimal numbers.

In addition to standard and number-and-word notation, we will learn new ways to represent large and small numbers using exponential and scientific notation. Your child will realize that scientific notation, which is used by scientists and mathematicians, is an easier and more efficient way to write large numbers. For example, the distance from the Sun to the planet Pluto is 3,675,000,000 miles. In scientific notation, the same number is expressed as $3.675 * 10^9$.

To use scientific notation, your child will first need to know more about exponential notation, which is a way of representing multiplication of repeated factors. For example, $7 * 7 * 7 * 7$ can be written as 7^4. Similarly, 100,000, or $10 * 10 * 10 * 10 * 10$, is also 10^5.

Unit 2 also reviews multiplication and division of whole numbers. All these strategies will be extended to decimals. The partial-quotient algorithm used in fourth and fifth grade *Everyday Mathematics* to divide whole numbers will also be used to divide decimals to obtain decimal quotients. This algorithm is similar to the traditional long division method, but it is easier to learn and apply. The quotient is built up in steps using "easy" multiples of the divisor. The student doesn't have to get the partial quotient exactly right at each step. The example below demonstrates how to use the partial-quotient algorithm.

Example:

Partial-Quotient Algorithm

```
12)  3270          Partial Quotients
   - 2400    200 ←— 200 * 12 = 2,400
     870          100 * 12 = 1,200
   -  600     50 ←— 50 * 12 = 600
     270          ⟋ 20 * 12 = 240
   -  240     20 ⟋  10 * 12 = 120
      30           5 * 12 = 60
   -   24      2 ←— 2 * 12 = 24
       6     272
       ↑      ↑
  Remainder  Quotient
```

The partial-quotient algorithm is discussed on pages 22 and 23 in the *Student Reference Book*.

Please keep this Family Letter for reference as your child works through Unit 2.

Vocabulary

Important terms in Unit 2:

dividend In division, the number that is being divided. For example, in $35 \div 5 = 7$, the dividend is 35.

$$\text{dividend} / \text{divisor} = \text{quotient}$$
$$\frac{\text{dividend}}{\text{divisor}} = \text{quotient}$$

divisor In division, the number that divides another number (the *dividend*). For example, in $35 / 5 = 7$, the divisor is 5.

exponent A small, raised number used in *exponential notation* to tell how many times the base is used as a *factor*. For example, in 5^3, the base is 5, the exponent is 3, and $5^3 = 5 * 5 * 5$. Same as *power*.

exponential notation A way of representing repeated multiplication by the same factor. For example, 2^3 is exponential notation for $2 * 2 * 2$. The *exponent* 3 tells how many times the base 2 is used as a factor.

factor (1) Each of two or more numbers in a product. For example, in $6 * 0.5$, 6 and 0.5 are factors. Compare to *factor of a counting number* n. (2) To represent a number as a product of factors. For example, factor 21 by rewriting as $7 * 3$.

number-and-word notation A notation consisting of the significant digits of a number and words for the place value. For example, 27 billion is number-and-word notation for 27,000,000,000.

power Same as *exponent*.

power of 10 (1) In *Everyday Mathematics*, a number that can be written in the form 10^a, where *a* is a counting number. That is, the numbers $10 = 10^1$, $100 = 10^2$, $1000 = 10^3$, and so on, that can be written using only 10s as factors. Same as positive power of 10. (2) More generally, a number that can be written in the form 10^a, where *a* is an integer. That is, all the positive and negative powers of 10 together, along with $10^0 = 1$.

precise Exact or accurate.

precise measures The smaller the scale of a measuring tool, the more *precise* a measurement can be. For example, a measurement to the nearest inch is more precise than a measurement to the nearest foot. A ruler with $\frac{1}{16}$-inch markings can be more precise than a ruler with only $\frac{1}{4}$-inch markings, depending on the skill of the person doing the measuring.

precise calculations The more accurate measures or other data are, the more *precise* any calculations using those numbers can be.

quotient The result of dividing one number by another number. For example, in $10 / 5 = 2$, the quotient is 2.

remainder An amount left over when one number is divided by another number. For example, in $16 / 3 \rightarrow 5$ R1, the quotient is 5 and the remainder R is 1.

scientific notation A way of writing a number as the product of a *power of 10* and a number that is at least 1 and less than 10. Scientific notation allows you to write large and small numbers with only a few symbols. For example, in scientific notation, 4,300,000 is $4.3 * 10^6$, and 0.00001 is 1×10^{-5}. Scientific calculators display numbers in scientific notation. Compare to *standard notation* and expanded notation.

standard notation Our most common way of representing whole numbers, integers, and decimals. Standard notation is base-ten place-value numeration. For example, standard notation for three hundred fifty-six is 356. Same as decimal notation.

Do-Anytime Activities

Consider using the suggested real-life applications and games that not only promote your child's understanding of Unit 2 concepts, but also are easy, fun, and rewarding to do at home.

1. Encourage your child to incorporate math vocabulary in everyday speech. Help your child recognize the everyday uses of fractions and decimals in science, statistics, business, sports, print and television journalism, and so on.

2. Have your child help you measure ingredients when cooking or baking at home. This will usually involve working with fractional amounts. Furthermore, your child could assist you with adjusting the amounts for doubling a recipe or making multiple servings.

3. Extend your child's thinking about fractions and decimals to making connections with percents. By using money as a reference, you could help your child recognize that one-tenth is equal to $\frac{10}{100}$ or 10%, one-quarter is the same as 0.25, $\frac{25}{100}$, or 25%, and so on.

4. Ask your child to use mental math skills to help you calculate tips. For example, if the subtotal is $25.00 and the tip you intend to pay is 15%, have your child first find 10% of $25 ($2.50) and then find 5% of $25 by taking half the 10% amount ($2.50 / 2 = $1.25). Add $2.50 and $1.25 to get the tip amount of $3.75.

Building Skills through Games

Several math games develop and reinforce whole number and decimal concepts in Unit 2. Detailed game instructions for all sixth-grade games are provided in the *Student Reference Book*. Encourage your child to play the following games with you at home.

Scientific Notation Toss See *Student Reference Book*, page 331.
Two players can play this game using a pair of 6-sided dice. Winning the game depends on creating the largest number possible using

scientific notation. *Advanced Scientific Notation Toss*, mentioned at the bottom of page 331, adds more excitement to the original game.

Doggone Decimal See *Student Reference Book*, page 310.
In this game, two players compete to collect the greatest number of cards. You will need number cards, 4 index cards, 2 counters or coins, and a calculator. The skill practiced here is estimating products of whole and decimal numbers.

As You Help Your Child with Homework

As your child brings assignments home, you might want to go over the instructions together, clarifying them as necessary. The answers listed below will guide you through the unit's Study Links.

Study Link 2•1

1. a. 2   **b.** 5   **c.** 1   **d.** 6   **e.** 8   **f.** 0

2. a. 430,000   **b.** 90,105,000
   c. 170,000,065   **d.** 9,500,243,000

3. a. (3 * 100,000) + (2 * 10,000) + (1 * 1,000)

4. a. 1,000   **b.** 1,000,000   **c.** 1,000,000,000

5. a. 48 million miles   **b.** 25.7 million miles

6. a. 44,300,000,000   **b.** 6,500,000,000,000
   c. 900,000   **d.** 70

7. 416,300   **8.** 230,000   **9.** 1,900,000

10. 7,000,000

Study Link 2•2

1. 38.469   **2.** 1.3406   **3.** eight-tenths

4. ninety-five hundredths   **5.** five-hundredths

7. four and eight hundred two ten-thousandths

11. (1 * 0.01) + (3 * 0.001)

12. (1 * 100) + (9 * 1) + (3 * 0.1) + (5 * 0.01) +
(2 * 0.001) + (7 * 0.0001)

13. 8.630   **14.** 0.368   **15.** *D*   **16.** *A*

17. *C*   **18.** *B*   **19.** 0.63   **20.** 0.0168

21. 0.7402   **22.** 45.009   **23.** 0.5801

Study Link 2•3

1. 0.297 minutes   **2.** 5.815 meters

3. 1.339 mph   **4.** 1.38 goals

7. $0.71   **8.** 0.85   **9.** 1.5   **10.** $6.75

Study Link 2•4

1. 0.0049   **2.** 0.078   **3.** 3.0   **4.** 0.07

5. 150.0   **6.** 190   **7.** 3,760   **8.** 0.0428

9. a. 100   **b.** 10^{100} **10.** 0.000000001

11. 10^7 **12.** $5.25   **13.** $6.02   **14.** $9.11

Study Link 2•5

1. 2,001   **2.** 1,288   **3.** 11,904

4. a. 20.01   **b.** 20.01   **c.** 200.1

5. a. 1,190.4   **b.** 11.904   **c.** 11.904

7. $5.00   **8.** $11.00   **9.** 34.5   **10.** 0.07

Study Link 2•6

1. 24.3   **2.** 11.48   **3.** 0.827   **4.** 756.3

5. 18.012   **6.** 29.82   **7.** 49.92   **8.** 10.241

9. 76.7 miles; 11.8 * 6.5 = 76.7

12. $16.00   **13.** $11.00   **14.** 96   **15.** 24

Study Link 2•7

6. → 66 R6; $66\frac{6}{8}$ **7.** → 65 R1; $65\frac{1}{15}$ **8.** = 49

9. →18 R15; $18\frac{15}{46}$ **10.** → 158 R20; $158\frac{20}{38}$

11. →126 R42; $126\frac{42}{44}$

12. $3.98   **13.** $11.84   **14.** $74.94   **15.** $499.95

Study Link 2•8

1. Sample estimate: 2; Answer: 2.47

2. Sample estimate: 20; Answer: 19.7

5. 2.83   **6.** $7.20   **7.** 1.99   **8.** 4.22

Study Link 2•9

1. 12,400   **3.** 0.000008   **5.** $1.1802 * 10^{10}$

6. 0.00016   **7.** $4.3 * 10^{-3}$ **8.** 2,835,000

9. >   **10.** =   **11.** <   **12.** >

13. 10 is raised to a negative power.

14. 7,624   **15.** 3.71   **16.** 900   **17.** 200

Study Link 2•10

1. 49   **3.** 64   **5.** 0.00001

7. 3^9 **9.** 11^{-3} **14.** 8^5 = 32,768

Study Link 2•11

1. $3.6 * 10^{-3}$ **3.** $8 * 10^4$ **5.** 50,000

7. 48,100,000   **9.** $1 * 10^{-3}$; 0.001   **11.** $3.9 * 10^3$

13. $5.2 * 10^{-1}$ **16.** 6,763 − 3,929 = 2,834

17. 71,146 − 4,876 = 66,270

32

STUDY LINK 2·1 Large Numbers

trillions			,	billions			,	millions			,	thousands			,	ones		
100,000,000,000,000	10,000,000,000,000	1,000,000,000,000		100,000,000,000	10,000,000,000	1,000,000,000		100,000,000	10,000,000	1,000,000		100,000	10,000	1,000		100	10	1

SRB
4

1. Write the digit in each place of the number 6,812,507,439.

   a. millions _____   b. hundred thousands _____   c. ten millions _____

   d. billions _____   e. hundred millions _____   f. ten thousands _____

2. Write each of the following numbers in standard form.

   a. four hundred thirty thousand _____

   b. ninety million, one hundred five thousand _____

   c. one hundred seventy million, sixty-five _____

   d. nine billion, five hundred million,
      two hundred forty-three thousand _____

3. Write each number in expanded form. **Example:** 235 = (2 * 100) + (3 * 10) + (5 * 1)

   a. 321,000

   _____

   b. 7,300,000,000,000

   _____

   c. 2,510,709

   _____

4. Use extended facts to complete the following.

   a. 1 million = 1,000 * _____

   b. 1 billion = 1,000 * _____

   c. 1 trillion = 1,000 * _____

33

STUDY LINK 2·1 | **Large Numbers** *continued*

___ ___ ___ , ___ ___ ___ , ___ ___ ___ , ___ ___ ___ , ___ ___ ___

trillion, *billion*, *million*, *thousand*,

Because the orbits of the planets are elliptical in shape, the distance between two planets changes over time. The least distances of Mercury, Venus, Saturn, and Neptune from Earth appear in the table at the right. The distances are approximations.

Least Distance from Earth	
Planet	**Distance (in miles)**
Mercury	48,000,000
Venus	25,700,000
Saturn	850,000,000
Neptune	2,680,000,000

5. Write each planet's least distance from Earth in number-and-word notation.

a. Mercury _____   **b.** Venus _____

c. Saturn _____   **d.** Neptune _____

6. Write the following numbers in standard notation.

a. 44.3 billion _____   **b.** 6.5 trillion _____

c. 0.9 million _____   **d.** 0.7 hundred _____

Practice

Round each number to the given place.

7. 416,254; hundreds

_____

8. 234,989; ten thousands

_____

9. 1,857,000; hundred thousands

_____

10. 6,593,278; millions

_____

Writing Decimals

26-28

1. Build a numeral. Write:
9 in the thousandths place,
4 in the tenths place,
8 in the ones place,
3 in the tens place, and
6 in the hundredths place.

Answer:

—— ——.—— —— ——

2. Build a numeral. Write:
3 in the tenths place,
6 in the ten-thousandths place,
4 in the hundredths place,
0 in the thousandths place, and
1 in the ones place.

Answer:

——.—— —— —— ——

Write the following numbers in words.

3. 0.8 _____

4. 0.95 _____

5. 0.05 _____

6. 0.067 _____

7. 4.0802 _____

Write a decimal place value in each blank space.

8. Bamboo grows at a rate of about 0.00004, or four _____,
kilometer per hour.

9. The average speed that a certain brand of catsup pours from the mouth of the bottle is
about 0.003, or three _____, mile per hour.

10. A three-toed sloth moves at a speed of about 0.068 to 0.098, or sixty-eight
_____ to ninety-eight _____, mile per hour.

Name	Date	Time

Writing Decimals *continued*

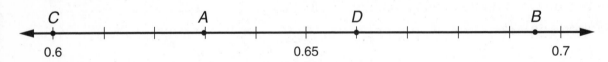

Write each of the following numbers in expanded notation.

Example: $2.756 = (2 * 1) + (7 * 0.1) + (5 * 0.01) + (6 * 0.001)$

11. 0.013 _____

12. 109.3527 _____

_____

13. Using the digits 0, 3, 6, and 8, write the greatest decimal number possible.

___.___ ___ ___

14. Using the digits 0, 3, 6, and 8, write the least decimal number possible.

___.___ ___ ___

Try This

C   A   D   B

0.6   0.65   0.7

Name the point on the number line that represents each of the following numbers.

15. 0.66 _____   **16.** 0.6299 _____   **17.** 0.6 _____   **18.** 0.695 _____

19. Refer to the number line above. Round 0.6299 to the nearest hundredth. _____

Practice

20. $0.01 + 0.006 + 0.0008 =$ _____

21. $0.7 + 0.04 + 0.0002 =$ _____

22. _____ $= 40 + 5 + 0.009$

23. _____ $= 0.50 + 0.080 + 0.00010$

36

STUDY LINK
2·3

Sports Records

SRB
31–33
257

Solve.

1. The fastest winning time for the New York Marathon (Tesfay Jifar of Ethiopia, 2001) is 2 hours, 7.72 minutes. The second fastest time is 2 hours, 8.017 minutes (Juma Ikangaa of Tanzania, 1989).

   How much faster was Jifar's time than Ikangaa's? _____

2. In the 1908 Olympic Games, Erik Lemming of Sweden won the javelin throw with a distance of 54.825 meters. He won again in 1912 with a distance of 60.64 meters.

   How much longer was his 1912 throw than his 1908 throw?

   _____

3. Driver Buddy Baker (Oldsmobile, 1980) holds the record for the fastest winning speed in the Daytona 500 at 177.602 miles per hour. Bill Elliott (Ford, 1987) has the second fastest speed at 176.263 miles per hour.

   How much faster is Baker's speed than Elliott's?

   _____

4. The highest scoring World Cup Soccer Final was in 1954. Teams played 26 games and scored 140 goals for an average of 5.38 goals per game. In 1950, teams played 22 games and scored 88 goals for an average of 4 goals per game.

   What is the difference between the 1954 and the 1950 average goals per game?

   _____

5. 46.09 + 123.047   Estimate _____

   46.09 + 123.047 = _____

6. 0.172 + 4.5   Estimate _____

   0.172 + 4.5 = _____

Practice

Solve mentally.

7. $0.36 + $0.29 + $0.64 + _____ = $2.00

8. 7.03 + _____ + 14.05 + 13.07 = 35

9. 9.225 + 8.5 + 5.775 + _____ = 25

10. $3.69 + _____ + $8.31 + $6.25 = $25

STUDY LINK 2·4 — Multiplying by Powers of 10

Some Powers of 10

10^4	10^3	10^2	10^1	10^0	.	10^{-1}	10^{-2}	10^{-3}	10^{-4}
10 * 10 * 10 * 10	10 * 10 * 10	10 * 10	10	1	.	$\frac{1}{10}$	$\frac{1}{10} * \frac{1}{10}$	$\frac{1}{10} * \frac{1}{10} * \frac{1}{10}$	$\frac{1}{10} * \frac{1}{10} * \frac{1}{10} * \frac{1}{10}$
10,000	1,000	100	10	1	.	0.1	0.01	0.001	0.0001

Multiply.

1. $4.9 * 0.001 =$ _____

2. _____ $= 7.8 * 0.01$

3. $30 * 10^{-1} =$ _____

4. _____ $= 7 * 10^{-2}$

5. $0.15 * 10^3 =$ _____

6. _____ $= 1.9 * 100$

7. $37.6 * 10^2 =$ _____

8. $42.8 * 10^{-3} =$ _____

9. Mathematician Edward Kasner asked his 9-year-old nephew to invent a name for the number represented by 10^{100}. The boy named it a *googol*. Later, an even larger number was named—a *googolplex.* This number is represented by 10^{googol}, or $10^{10^{100}}$.

   a. How many zeros are in the standard form of a googol, or 10^{100}? _____

   b. One googolplex is 1 followed by how many zeros? _____

10. The speed of computer memory and logic chips is measured in nanoseconds. A nanosecond is one-billionth of a second, or 10^{-9} second. Write this number in standard form. _____

11. Light travels about 1 mile in 0.000005 seconds. If a spacecraft could travel at this speed, it would travel almost 10^6 miles in 5 seconds. About how far would this spacecraft travel in 50 seconds? _____ miles

Practice

Mentally calculate your change from $10.

12. Cost: $4.75; Change: _____

13. Cost: $3.98; Change: _____

14. Cost: $0.89; Change: _____

15. Cost: $8.46; Change: _____

STUDY LINK 2·5 Multiplying Decimals: Part 1

Multiply.

1.   23
       * 87
    _____

2.   56
       * 23
    _____

3.   124
       * 96
    _____

4. Use your answer for Problem 1 to place the decimal point in each product.

  a. 2.3 * 8.7 = _____

  b. 23 * 0.87 = _____

  c. 2.3 * 87 = _____

5. Use your answer for Problem 3 to place the decimal point in each product.

  a. 124 * 9.6 = _____

  b. 1.24 * 9.6 = _____

  c. 12.4 * 0.96 = _____

Two new U.S. nickels were issued in 2004. A likeness of Thomas Jefferson remained on the front of the nickels. The reverse side featured images commemorating either the Louisiana Purchase or the Lewis and Clark expedition.

6. A U.S. nickel is 1.95 mm thick.

  a. Estimate the height of a stack of 25 nickels.   Estimate _____ mm

  b. Calculate the actual height of the stack in mm. _____ mm

  c. How much is a stack of 25 nickels worth? _____

Practice

Multiply by 0.10 to find 10% of each number.

7. 10% of $50.00 = _____

8. 10% of $110.00 = _____

9. 10% of 345 = _____

10. 10% of 0.70 = _____

STUDY LINK 2·6 | Multiplying Decimals: Part 2

SRB
37-39

Place a decimal point in each problem.

1. 2 4 3 * 7.06 = 171.558

2. 16.4 * 0.7 = 1 1 4 8

3. 8 2 7 * 9.5 = 7.8565

4. 7 5 6 3 * 5.1 = 3,857.13

Multiply. Show your work on a separate sheet of paper or on the back of this page.

5. _____ = 2.28 * 7.9

6. _____ = 49.7 * 0.6

7. _____ = 3.84 * 13

8. _____ = 0.19 * 53.9

Solve each problem. Then write a number model.
(*Hint: Change fractions to decimals.*)

9. Janine rides her bike at an average speed
   of 11.8 miles per hour. At that speed, about
   how many miles can she ride in $6\frac{1}{2}$ hours? _____

   Number Model _____

10. Kate types at an average rate of 1.25 pages
    per quarter hour. If she types for $2\frac{3}{4}$ hours,
    about how many pages can she type? _____

    Number Model _____

11. Find the area in square meters of a
    rectangle with length 1.4 m and width 2.9 m. _____

    Number Model _____

Practice

Multiply mentally by 0.10 to find 10%. Then mentally calculate the percent that
has been assigned to each number.

12. 20% of $80.00 = _____

13. 5% of $220.00 = _____

14. 15% of 640 = _____

15. 30% of 80 = _____

STUDY LINK 2·7 **Dividing Numbers**

3 Ways to Write a Division Problem

246 ÷ 12 → 20 R6   12)246 → 20 R6   246 / 12 → 20 R6

2 Ways to Express a Remainder

12)246 → 20 R6   12)246 = $20\frac{6}{12}$, or $20\frac{1}{2}$

When estimating quotients, use "close" numbers that are easy to divide.

Example: 346 / 12   Estimate ___*35*___   How I estimated: *350 / 10 = 35*

1. 234 / 6   Estimate _____   How I estimated: _____

2. 659 / 12   Estimate _____   How I estimated: _____

3. 512 / 9   Estimate _____   How I estimated: _____

4. 1,270 / 7   Estimate _____   How I estimated: _____

5. 728 / 34   Estimate _____   How I estimated: _____

Solve using a division algorithm. Show your work on a separate sheet of paper or a computation grid.

6. 8)534 _____   **7.** 976 / 15 _____

8. 980 ÷ 20 _____   **9.** 46)843 _____

10. 6,024 / 38 _____   **11.** 5,586 ÷ 44 _____

Practice

Multiply mentally.

12. 2 notebooks at $1.99 each = _____

13. 4 pens at $2.96 each = _____

14. 3 books at $24.98 each = _____

15. 5 gifts at $99.99 each = _____

STUDY LINK 2·8 Dividing Decimals

For each problem, follow the steps below. Show your work on a separate sheet of paper or a computation grid.

◆ Estimate the quotient. Use numbers that are close to the numbers given and that are easy to divide. Write your estimate. Then write a number sentence to show how you estimated.

◆ Ignore any decimal points. Divide as if the numbers were whole numbers.

◆ Use your estimate to insert a decimal point in the final answer.

1. 19.76 ÷ 8   Estimate _____

_____
How I estimated

Answer _____

2. 78.8 / 4   Estimate _____

_____
How I estimated

Answer _____

3. 85.8 / 13   Estimate _____

_____
How I estimated

Answer _____

4. 51.8 / 7   Estimate _____

_____
How I estimated

Answer _____

5. Find 17 ÷ 6. Give the answer as a decimal with 2 digits after the decimal point.

_____

6. Five people sent a $36 arrangement of flowers to a friend. Divide $36 into 5 equal shares. How much is 1 share, in dollars and cents?

_____

Practice

Divide mentally to find the price for 1 pound (lb).

7. $3.98 for 2 lb = $_____ per 1 lb

8. $16.88 for 4 lb = $_____ per 1 lb

9. $45.80 for 5 lb = $_____ per 1 lb

10. $299.10 for 10 lb = $_____ per 1 lb

STUDY LINK 2·9 Using Scientific Notation

Write each number in standard notation.

1. $1.24 * 10^4 =$ _____   **2.** $3.5 * 10^{-3} =$ _____

3. $8 * 10^{-6} =$ _____   **4.** $7.061 * 10^8 =$ _____

Change the numbers given in standard notation to scientific notation. Change the numbers given in scientific notation to standard notation.

5. Light travels about 11,802,000,000, or _____, inches per second.

6. A bacterium can travel across a table at a speed of $1.6 * 10^{-4}$,

or _____, km per hour.

7. One dollar bill has a thickness of 0.0043, or _____, inches.

8. The mass of 1 million pennies is approximately $2.835 * 10^6$,

or _____, grams.

Use $<$, $>$, or $=$ to compare each pair of numbers.

9. 10^{-2} _____ 10^{-3} **10.** $1.23 * 10^{-3}$ _____ $\dfrac{1.23}{1,000}$

11. $9.87 * 10^5$ _____ $1.2 * 10^6$ **12.** $5.4 * 10^{-1}$ _____ $9.6 * 10^{-4}$

13. Explain how you can tell whether a number written in scientific notation is less than 1.

_____

_____

Practice

Solve mentally.

14. $3,625 + 3,999 =$ _____   **15.** $8.7 - 4.99 =$ _____   **16.** $4 * 225 =$ _____

17. $100,000 / 500 =$ _____   **18.** $683 - 298 =$ _____   **19.** $387 + 499 =$ _____

STUDY LINK 2·10 | **Exponential Notation**

Use your calculator to write each number in standard notation.

1. $7^2 =$ _____

2. $(0.25)^2 =$ _____

3. $4^3 =$ _____

4. $(0.41)^3 =$ _____

5. $10^{-5} =$ _____

6. $(2.5)^{-3} =$ _____

Use digits to write each number in exponential notation.

7. three to the ninth power _____

8. eight to the seventh power _____

9. eleven to the negative third power _____

10. five-tenths to the negative sixth power _____

Write each number as a product of repeated factors.

Example: $5^3 = 5 * 5 * 5$

11. $(\frac{1}{2})^5 =$ _____

12. $10^{-2} =$ _____

13. $10^{-6} =$ _____

14. You can find the total number of different 4-digit numbers that can be made using the digits 1 through 9 by raising the number of choices for each digit (9) to the number of digits (4), or 9^4.

Based on this pattern, how many different 5-digit numbers could you make from the digits 1 through 8? _____

Practice

Solve mentally.

15. $15.32 - 1.88 =$ _____

16. $7,200 / 90 =$ _____

17. $4.98 + 3.99 =$ _____

18. $8 * 525 =$ _____

STUDY LINK 2·11 Scientific Notation

Write the following numbers in scientific notation.

1. 0.0036 _____

2. 0.0007 _____

3. 80,000 _____

4. 600 thousand _____

Write the following numbers in standard notation.

5. $5 * 10^4$ _____

6. $4.73 * 10^9$ _____

7. $4.81 * 10^7$ _____

8. $8.04 * 10^{-2}$ _____

Write the next two numbers in each pattern.

9. $1 * 10^{-1}$; 0.1; $1 * 10^{-2}$; 0.01; _____ ; _____

10. 0.01, 0.002, 0.0003, _____ , _____

Solve the following problems. Write each answer in scientific notation.

11. $(4 * 10^3) - 10^2 =$ _____

12. $10^3 - (2 * 10^1) =$ _____

13. $(5 * 10^{-1}) + 0.02 =$ _____

14. $(7 * 10^4) - 10^3 =$ _____

15. Use a calculator to complete the table.

Problem	Calculator Display	Scientific Notation	Standard Notation
$5,000,000^2$			
$90^4 - 300^2$			
$20^3 + 30^2$			
$10^4 * 10^4$			
$5^{20} / 5^{16}$			

Practice

Find the missing digits to complete each number sentence.

16. ☐,☐63 − 3,9☐9 = 2,83☐

17. 71,☐4☐ − 4,8☐6 = 6☐,270

STUDY LINK 2·12 | **Unit 3: Family Letter**

Variables, Formulas, and Graphs

In Unit 3, students will be introduced to variables—symbols such as *x*, *y*, and *m*—that stand for a specific number or any number in a range of values. The authors of *Everyday Mathematics* believe that work with variables is too important to be delayed until high-school algebra courses. The problem "Solve $3x + 40 = 52$" might be difficult for some high-school students because they see it as merely symbol manipulation. Problems such as these are posed to *Everyday Mathematics* students as puzzles that can be unraveled by asking, "What number makes the equation true?" *I need to add 12 to 40 to get 52. Three times what number yields 12? The answer is* x = 4.

In addition to being used in algebraic equations, variables are also used to describe general patterns, to form expressions that show relationships, and to write rules and formulas. Unit 3 will focus on these three uses of variables.

In this unit, your child will work with "What's My Rule?" tables like the one below (introduced in early grades of *Everyday Mathematics*). He or she will learn to complete such tables following rules described in words or by algebraic expressions. Your child will also determine rules or formulas from information given in tables and graphs.

Rule: $y = (4 * x) + -3$

x	y
5	17
2	
0	
	37

In addition, your child will learn how to name cells in a spreadsheet and write formulas to express the relationships among spreadsheet cells. If you use computer spreadsheets at work or at home, you may want to share your experiences with your child. The class will play *Spreadsheet Scramble,* in which students practice computation and mental addition of positive and negative numbers. Encourage your child to play a game at home. See the *Practice through Games* section of this letter for some suggestions.

	A	B	C	D	E	F
1						**Total**
2						
3						
4						
5	**Total**					

Please keep this Family Letter for reference as your child works through Unit 3.

Math Tools

Your child will be using **spreadsheets**, a common mathematics tool for the computer. The spreadsheet, similar to the one shown here, gets its name from a ledger sheet for financial records. Such sheets were often large pages, folded or taped, that were *spread* out for examination.

	A	B	C	D
		Class picnic ($$)		
1		budget for class picnic		
2				
3	quantity	food items	unit price	cost
4	6	packages of hamburgers	2.79	16.74
5	5	packages of hamburger buns	1.29	6.45
6	3	bags of potato chips	3.12	9.36
7	3	quarts of macaroni salad	4.50	13.50
8	4	bottles of soft drinks	1.69	6.76
9			subtotal	52.81
10			8% tax	4.23
11			total	57.04

Vocabulary

Important terms in Unit 3:

algebraic expression An expression that contains a variable. For example, if Maria is 2 inches taller than Joe, and if the variable *m* represents Maria's height, then the algebraic expression $m - 2$ represents Joe's height.

cell In a spreadsheet, a box formed where a column and a row intersect. A *column* is a section of cells lined up vertically. A *row* is a section of cells lined up horizontally.

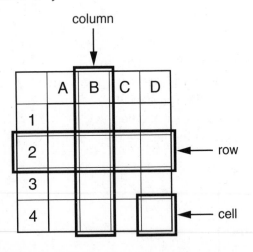

formula A general rule for finding the value of something. A formula is often written using letters, called *variables,* that stand for the quantities involved. For example, the formula for the area of a rectangle may be written as $A = b * h$, where *A* represents the area of the rectangle, *b* represents its base, and *h* represents its height.

general pattern In *Everyday Mathematics,* a number model for a pattern or rule.

special case In *Everyday Mathematics,* a specific example of a *general pattern.* For example, $6 + 6 = 12$ is a special case of $y + y = 2y$ and $9 = 4.5 * 2$ is a special case of $A = l * w$. Same as instance of a pattern.

time graph A graph representing a story that takes place over time. For example, the time graph below shows the trip Mr. Olds took to drive his son to school. The line shows the increases, decreases, and constant rates of speed that Mr. Olds experienced during the 13-minute trip.

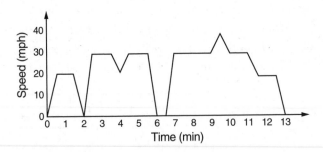

variable A letter or symbol that represents a number. A variable can represent one specific number, or it can stand for many different numbers.

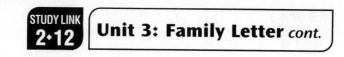

Do-Anytime Activities

Try these ideas to help your child with the concepts taught in this unit.

1. If you are planning to paint or carpet a room, consider having your child measure and calculate the area using the area formula for rectangular surfaces: Area = base * height. If the room is irregular in shape, divide it into rectangular regions, find the area of each region, and add all the areas to find the total area. If a room has a cathedral ceiling, imagine a line across the top of the wall to form a triangle. Your child can use the area formula for triangles: Area = $\frac{1}{2}$ * (base * height), to calculate the area of the triangle.

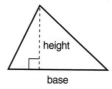

2. If you use a spreadsheet program on a home computer, help your child learn how to use it. You might help your child set up a spreadsheet to keep track of his or her math scores and to figure out the mean.

3. Practice renaming fractions, which is a prerequisite skill for Unit 4. **Examples:**

Rename as Fractions

$3\frac{1}{2} = \underline{\quad \frac{7}{2} \quad}$

$8\frac{1}{3} = \underline{\quad \frac{25}{3} \quad}$

Rename as Mixed or Whole Numbers

$\frac{33}{5} = \underline{\quad 6\frac{3}{5} \quad}$

$\frac{25}{5} = \underline{\quad 5 \quad}$

Building Skills through Games

The concepts learned in Unit 3 will be reinforced through several math games included in this unit that are fun to play in class and at home. Detailed game instructions for all sixth-grade games are available in the games section of the *Student Reference Book.* Here is a list and a brief description of some of the games in this unit:

Getting to One See *Student Reference Book,* page 321
Two players can play this game using a calculator. The object of the game is to divide a number by a mystery number and to find the mystery number in as few tries as possible. Players apply place-value concepts of decimal numbers to determine which numbers to play.

Division Top-It (**Advanced Version**) See *Student Reference Book,* page 336
Two to four people can play this game using number cards 1 through 9. Players apply place-value concepts, division facts, and estimation strategies to generate whole-number division problems that will yield the largest quotient.

As You Help Your Child with Homework

As your child brings assignments home, you may want to go over the instructions together, clarifying them as necessary. Some of the answers listed below will guide you through the unit's Study Links.

Study Link 3·1

Sample answers (1–7):

1. **a.** The sum of any number and 0 is equal to the original number.
   b. $36.09 + 0 = 36.09$; $52 + 0 = 52$

2. $(2 * 24) + 24 = 3 * 24$; $(2 * 10) + 10 = 3 * 10$

3. $100 + 0.25 = 0.25 + 100$;
   $0.5 + 0.25 = 0.25 + 0.5$

4. $x^2 * x^3 = x^5$ 5. $s * 0.1 = \frac{s}{10}$ 6. $m^0 = 1$

7. 10   8. 100, 0.25   9. 20

10. 75, 100   11. 80, 0.80   12. 70, 0.70

Study Link 3·2

Sample answers (1–7):

1. $(6 * 2) * 3 = 6 * (2 * 3)$;
   $(6 * 1) * 5 = 6 * (1 * 5)$

2. $12 \div (\frac{6}{2}) = (2 * 12) \div 6$; $10 \div (\frac{4}{2}) = (2 * 10) \div 4$

3. $\frac{10}{5} = 10 * \frac{1}{5}$, $\frac{3}{4} = 3 * \frac{1}{4}$

4. $a - b = a + (-b)$ 5. $\frac{m}{n} = \frac{m*3}{n*3}$

6. $\frac{s}{t} = \frac{s \div 2}{t \div 2}$ 7. $\frac{c}{d} * \frac{1}{2} = \frac{c*1}{d*2}$

8. 2.5   9. 1.06   10. 1.00

Study Link 3·3

1. $x - 7$ 2. $d + 2.5$ 3. $\frac{c}{12}$, $c \div 12$, or $\frac{1}{12}c$

4. $2 * h$, or $2h$; 8   5. $3r + 8$, or $(3 * r) + 8$; 44

6. 275   7. 35   8. 0.5

Study Link 3·4

1. **a.** Subtract 0.22 from m.
   b. $n = m - 0.22$

2. **a.** Multiply r by $\frac{1}{2}$ or divide r by 2.
   b. $r * 0.5 = t$

3. $q = (2 * p) - 2$

4. 15   5. 210   6. 1,760   7. 29,040

Study Link 3·5

1. *in*: $6\frac{1}{2}$; *out*: $10\frac{1}{2}$, $9\frac{1}{2}$, 3, $-\frac{1}{2}$

2. *in*: 6, $\frac{1}{4}$; *out*: 48, 1.2, 2   3. *in*: 7, 0; *out*: 0, 18

4. Divide the *in* number by 3; $d = b \div 3$

5. Answers vary.   6. –3   7. –12   8. –3   9. 10

Study Link 3·6

1. Perimeter (in.): 4, 8, 12, 16, 20;
   Area (in.²): 1, 4, 9, 16, 25

3. 10 in.   4. 17 in.   6. $2\frac{1}{4}$ in.²

7. $10\frac{1}{2}$ in.²   8. 54.45   9. 4.2

Study Link 3·7

1. January   2. $115.95   3. A5   4. C3

5. Column E: $118.75; $152.95; $2,625.00

6. E3 = B3 + C3 + D3   7. E5 = B5 + C5 + D5

8. $128.75   9. 144   10. 9   11. 73.96   12. 17

Study Link 3·8

1. –6   2. 4   3. 1   4. –6   5. 8

6. 2   7. –15   8. –5   9. –13   10. –12

11. 0, –2, –9
   a. Sample answer: Add –6 to x.
   b. $x + (-6) = y$

12. **a.** 25   **b.** 32   **c.** 50   **d.** –19

13. **a.** $\frac{1}{10}$ **b.** $\frac{1}{2}$ **c.** 2   **d.** –9

Study Link 3·9

1. Sample answer: People are getting on the Ferris wheel.

2. 125 sec   3. 170 sec   4. 4 times   5. 40 sec

Study Link 3·10

1. Jenna's Profit: $3, $6, $9, $12, $15;
   Thomas's Profit: $6, $8, $10, $12, $14

2. $18, $16   3. Jenna   4. Jenna's

5. $3, $2   6. (4,12)

7. **a.** 81   **b.** 8,000   **c.** 76   **d.** 875   **e.** 3

**STUDY LINK
3·1** **Variables in Number Patterns**

1. Following are 3 special cases representing a general pattern.

   $17 + 0 = 17$ $\qquad$ $-43 + 0 = -43$ $\qquad$ $\frac{7}{8} + 0 = \frac{7}{8}$

   a. Describe the general pattern in words.

   _____

   _____

   b. Give 2 other special cases for the pattern.

   _____   _____

For each general pattern, give 2 special cases.

2. $(2 * m) + m = 3 * m$

   _____   _____

3. $s + 0.25 = 0.25 + s$

   _____   _____

For each set of special cases, write a general pattern.

4. $3^2 * 3^3 = 3^5$

   $5^2 * 5^3 = 5^5$

   $13^2 * 13^3 = 13^5$

   _____

5. $7 * 0.1 = \frac{7}{10}$

   $3 * 0.1 = \frac{3}{10}$

   $4 * 0.1 = \frac{4}{10}$

   _____

6. $2^0 = 1$

   $146^0 = 1$

   $\left(\frac{1}{2}\right)^0 = 1$

   _____

Practice

Complete.

7. $\frac{1}{10} = \frac{\boxed{}}{100} = 0.10$

8. $\frac{1}{4} = \frac{25}{\boxed{}} = 0.\boxed{}$

9. $\frac{1}{5} = \frac{\boxed{}}{100} = 0.20$

10. $\frac{3}{4} = \frac{\boxed{}}{\boxed{}} = 0.75$

11. $\frac{4}{5} = \frac{\boxed{}}{100} = 0.\boxed{}$

12. $\frac{7}{10} = \frac{\boxed{}}{100} = 0.\boxed{}$

59

STUDY LINK 3·2 | **General Patterns with Two Variables**

For each general pattern, write 2 special cases.

1. $(6 * b) * c = 6 * (b * c)$ _____

_____

2. $a \div \dfrac{b}{2} = (2 * a) \div b$ _____

_____

3. $\dfrac{x}{y} = x * \dfrac{1}{y}$ _____

(y is not 0) _____

For each set of special cases, write a number sentence with 2 variables to describe the general pattern.

4. $7 - 5 = 7 + (-5)$

$12 - 8 = 12 + (-8)$

$9 - 1 = 9 + (-1)$

General pattern:

_____

5. $\dfrac{4}{6} = \dfrac{4 * 3}{6 * 3}$

$\dfrac{1}{2} = \dfrac{1 * 3}{2 * 3}$

$\dfrac{2}{5} = \dfrac{2 * 3}{5 * 3}$

General pattern:

_____

6. $\dfrac{6}{10} = \dfrac{6 \div 2}{10 \div 2}$

$\dfrac{4}{12} = \dfrac{4 \div 2}{12 \div 2}$

$\dfrac{2}{4} = \dfrac{2 \div 2}{4 \div 2}$

General pattern:

_____

7. $\dfrac{1}{5} * \dfrac{1}{2} = \dfrac{1 * 1}{5 * 2}$

$\dfrac{2}{3} * \dfrac{1}{2} = \dfrac{2 * 1}{3 * 2}$

$\dfrac{3}{4} * \dfrac{1}{2} = \dfrac{3 * 1}{4 * 2}$

General pattern:

_____

Practice

Write each fraction as a decimal.

8. $\dfrac{250}{100} =$ _____

9. $\dfrac{106}{100} =$ _____

10. $\dfrac{100}{100} =$ _____

61

**STUDY LINK
3·3**
General Patterns with Two Variables

Write an algebraic expression for each situation. Use the suggested variable.

1. Kayla has x CDs in her music collection. If Miriam has
7 fewer CDs than Kayla, how many CDs does Miriam have? _____ CDs

2. Chaz ran 2.5 miles more than Nigel. If Nigel ran
d miles, how far did Chaz run? _____ miles

3. If a car dealer sells c automobiles each year, what is the average
number of automobiles sold each month?

_____ automobiles

First translate each situation from words into an algebraic expression.
Then solve the problem that follows.

4. The base of a rectangle is twice the length of the height. If the
height of the rectangle is h inches, what is the length of the base?

_____ inches

If the height of the rectangle is 4 inches, what is the length of the base?

_____ inches

Try This

5. Monica has 8 more than 3 times the number of marbles Regina has. If Regina
has r marbles, how many marbles does Monica have?

_____ marbles

If Regina has 12 marbles, how many does Monica have? _____ marbles

---

Practice

6. 2.75 m = _____ cm   **7.** 3.5 cm = _____ mm   **8.** 500 m = _____ km

"What's My Rule?" Part 1

1. a. State in words the rule for the "What's My Rule?" table at the right.

   _____

   b. Which formula describes the rule? Fill in the circle next to the best answer.

   Ⓐ $n = m - 0.22$ Ⓑ $m + n = 0.22$ Ⓒ $m = n - 0.22$

m	n
4.56	4.34
10	9.78
0.01	−0.21
$\frac{24}{100}$	0.02
7.80	7.58

2. a. State in words the rule for the "What's My Rule?" table at the right.

   _____

   b. Which formula describes the rule? Fill in the circle next to the best answer.

   Ⓐ $r - 0.25 = t$ Ⓑ $t + 0.12 = r$ Ⓒ $r * 0.5 = t$

r	t
20	10
15	7.5
1	0.5
1.5	0.75
3.4	1.7

3. Which formula describes the rule for the "What's My Rule?" table at the right? Fill in the circle next to the best answer.

   Ⓐ $q - 13 = p$ Ⓑ $q = (2 * p) - 2$ Ⓒ $q = 2 * (p - 2)$

p	q
7	12
10	18
1	0
15	28
30	58

Practice

4. 180 in. = _____ feet

5. $3\frac{1}{2}$ minutes = _____ seconds

6. 5,280 ft = _____ yards

7. $5\frac{1}{2}$ miles = _____ feet

STUDY LINK
3·5

"What's My Rule?" Part 2

SRB
253

1. *Rule:* Subtract the *in* number from $11\frac{1}{2}$.

in	out
n	$11\frac{1}{2} - n$
1	
2	
$8\frac{1}{2}$	
	5
12	

2. *Formula:* $r = 4 * s$

in	out
s	r
12	
	24
0.3	
	1
$\frac{1}{2}$	

3. *Rule:* Triple the *in* number and add −6.

in	out
x	$(3x) + (-6)$
1	−3
2	
	15
8	
	−6

4. For the table below, write the rule in words and as a formula.

Rule: _____

_____

Formula: _____

in	out
b	d
1.5	0.5
$6\frac{3}{4}$	$2\frac{1}{4}$
24	8
81	27
9.75	3.25

5. Make up your own.

Rule: _____

_____

Formula: _____

in	out
x	y

Practice

6. $3 + -6 =$ _____

7. $-17 + 5 =$ _____

8. $8 + (-2) + (-9) =$ _____

9. $5 + 3 + (-5) + 7 =$ _____

67

STUDY LINK 3·6 Area and Perimeter

Perimeter

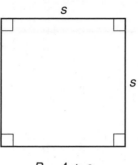

$P = 4 * s$

Area

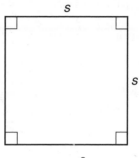

$A = s^2$

1. Use the perimeter and area formulas for squares to complete the table.

Length of side (in.)	Perimeter (in.)	Area (in.^{2})
1		
2		
3		
4		
5		

Use the table above to complete the graphs on *Math Masters,* page 87.

STUDY LINK
3·6

Area and Perimeter *continued*

2. Graph the perimeter data from the table on page 86.
Use the grid at the right.

Use the graph you made in Problem 2 to answer the
following questions.

3. If the length of the side of a square is $2\frac{1}{2}$ inches,
what is the perimeter of the square?

_____
(unit)

4. If the length of the side of a square is $4\frac{1}{4}$ inches,
what is the perimeter of the square?

_____
(unit)

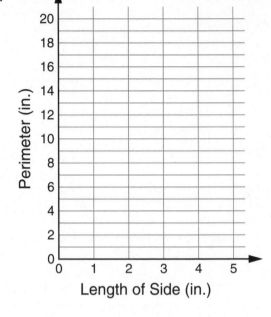

5. Graph the area data from the table on page 86.
Use the grid at the right.

Use the graph you made in Problem 5 to answer the
following questions.

6. If the length of the side of a square is $1\frac{1}{2}$ inches, what is
the approximate area of the square?

About _____
(unit)

7. If the length of the side of a square is $3\frac{1}{4}$ inches, what is
the approximate area of the square?

About _____
(unit)

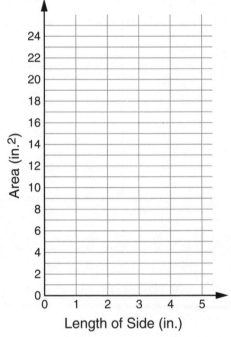

Practice

Find the missing dimension for each rectangle.

8. $b = 5.5$ cm; $h = 9.9$ cm; $A =$ _____ cm²

9. $b = 36$ in.; $h =$ _____ in.; $A = 151.2$ in.²

STUDY LINK 3·7 Spreadsheet Practice

Ms. Villanova keeps a spreadsheet of her monthly expenses. Use her spreadsheet to answer the questions below.

	A	B	C	D	E
1		January	February	March	**Total**
2	Groceries	$125.25	$98.00	$138.80	$362.05
3	Phone Bill	$34.90	$58.50	$25.35	
4	Car Expenses	$25.00	$115.95	$12.00	
5	Rent	$875.00	$875.00	$875.00	

1. What is shown in cell B1? _____

2. What is shown in cell C4? _____

3. Which cell contains the word *Rent*? _____

4. Which cell contains the amount $58.50? _____

5. Ms. Villanova used column E to show the total for each row. Find the missing totals and enter them on the spreadsheet.

6. Write a formula for calculating E3 that uses cell names. _____

7. Write a formula for calculating E5 that uses cell names. _____

8. Ms. Villanova found that she made a mistake in recording her March phone bill. Instead of $25.35, she should have entered $35.35. After she corrects her spreadsheet, what will the new total be in cell E3?

   _____

Practice

Find the missing dimension for each square.

9. $s = 12$ cm; $A =$ _____ cm²   10. $s =$ _____ in.; $A = 81$ in.²

11. $s = 8.6$ mm; $A =$ _____ mm²   12. $s =$ _____ ft; $A = 289$ ft²

STUDY LINK 3·8 Adding Positive and Negative Numbers

Solve.

1. $b + 9 = 3$; $b =$ _____

2. $-5 + a = -1$; $a =$ _____

3. $m + (-5) = -4$; $m =$ _____

4. $k + 3 = -3$; $k =$ _____

Add.

5. $13 + (-5) =$ _____

6. $(-10) + 12 =$ _____

7. _____ $= (-7) + (-8)$

8. _____ $= (-15) + 10$

9. $(-4) + (-9) =$ _____

10. _____ $= 7 + (-19)$

11. Complete the "What's My Rule?" table.

x	y
8	2
4	-2
2	-4
	-6
	-8
	-15

a. Give the rule for the table in words.

_____

b. Circle the formula that describes the rule.

$x + 6 = y$ $x * (-6) = y$ $x + (-6) = y$ $\frac{x}{6} = y$

Practice

12. Evaluate when $k = 5$.

a. k^2 _____

b. 2^k _____

c. $10k$ _____

d. $-24 + k$ _____

13. Evaluate when $x = -1$.

a. 10^x _____

b. 2^x _____

c. $(\frac{1}{2})^x$ _____

d. $x + (-8)$ _____

STUDY LINK 3·9 Ferris Wheel Time Graph

The time graph below shows the height of Rose's head from the ground as she rides a Ferris wheel. Use the graph to answer the following questions.

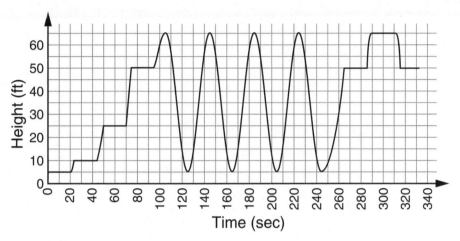

1. Explain what is happening from 0 to 95 seconds. _____

_____

2. How long is Rose on the Ferris wheel before she
   is back to the position from which she started?   About _____
                                                                (unit)

3. After the Ferris wheel has been completely loaded, about
   how long does the ride last before unloading begins?   _____
                                                                (unit)

4. After the Ferris wheel has been loaded, how many times
   does the wheel go around before unloading begins?   _____
                                                                (unit)

5. When the ride is in full swing, approximately how long
   does one complete revolution of the wheel take?   _____
                                                                (unit)

Try This

6. Rose takes another ride.
   After 130 seconds, the
   Ferris wheel comes to a
   complete stop because of
   an electrical failure. It starts
   moving again 2 minutes
   later. Complete the graph
   to show this event.

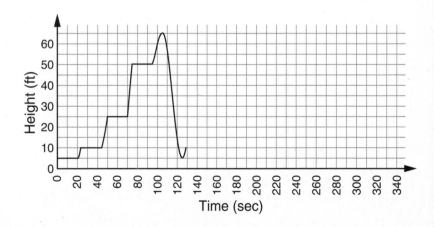

75

STUDY LINK 3·10 | **Comparing Pet-Sitting Profits**

Jenna and Thomas like to pet-sit for their neighbors. Jenna charges $3 per hour. Thomas charges $6.00 for the first hour and $2 for each additional hour.

1. Complete the table below. Use the table to graph the profit values for each sitter.

Time (hours)	Jenna's Profit ($)	Thomas's Profit ($)
1		
2		
3		
4		
5		

2. Extend both line graphs to find the profit each sitter will make for 6 hours.

Jenna (6 hours) _____   Thomas (6 hours) _____

3. Which sitter, Jenna or Thomas, earns more money for jobs of 5 hours or more? _____

4. Which line graph rises more quickly? _____

5. Complete each statement. For every hour that passes, Jenna's profit increases by _____; Thomas's profit increases by _____.

6. At what point do the line graphs intersect?

_____

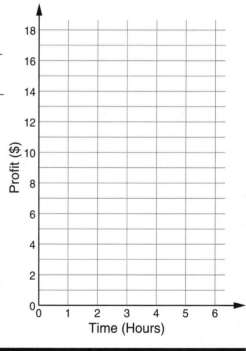

Practice

7. Evaluate when $m = 3$.

   a. m^4 _____ **b.** 20^m _____ **c.** $4^m + 4m$ _____ **d.** $10^m - 5^m$ _____ **e.** $\dfrac{m^3}{m^2}$ _____

Unit 4: Family Letter

Rational Number Uses and Operations

One reason for studying mathematics is that numbers in all their forms are an important part of our everyday lives. We use decimals when we are dealing with measures and money, and we use fractions and percents to describe parts of things.

Students using *Everyday Mathematics* began working with fractions in the primary grades. In *Fifth Grade Everyday Mathematics,* your child worked with equivalent fractions, operations with fractions, and conversions between fractions, decimals, and percents.

In Unit 4, your child will revisit these concepts and apply them. Most of the fractions with which your child will work (halves, thirds, fourths, sixths, eighths, tenths, and sixteenths) will be fractions that they would come across in everyday situations—interpreting scale drawings, following a recipe, measuring distance and area, expressing time in fractions of hours, and so on.

Students will be exploring methods for solving addition and subtraction problems with fractions and mixed numbers. They will look at estimation strategies, mental computation methods, paper-and-pencil algorithms, and calculator procedures.

Students will also work with multiplication of fractions and mixed numbers. Generally, verbal cues are a poor guide as to which operation $(+, -, *, /)$ to use when solving a problem. For example, *more* does not necessarily imply addition. However, *many of* and *part of* generally involve multiplication. At this point in the curriculum, your child will benefit from reading and understanding $\frac{1}{2} * 12$ as *one-half of 12,* rather than *one-half times 12;* or reading and understanding $\frac{1}{2} * \frac{1}{2}$ as *one-half of one-half,* rather than *one-half times one-half.*

Finally, students will use percents to make circle graphs to display the results of surveys and to learn about sales and discounts.

Jambalaya Recipe

4 ounces each of chicken and sausage

4 cups peppers

$\frac{3}{4}$ cup rice

$1\frac{2}{3}$ cups chopped onions

$1\frac{1}{2}$ tablespoons chopped thyme

$\frac{1}{8}$ teaspoon salt

Please keep this Family Letter for reference as your child works through Unit 4.

Math Tools

The **Percent Circle**, on the Geometry Template, is used to find the percent represented by each part of a circle graph and to make circle graphs. The Percent Circle is similar to a full-circle protractor with the circumference marked in percents rather than degrees. This tool allows students to interpret and make circle graphs before they are ready for the complex calculations needed to make circle graphs with a protractor.

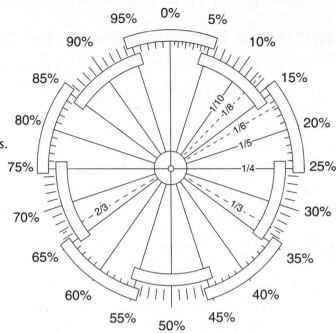

Vocabulary

Important terms in Unit 4:

common denominator A nonzero number that is a multiple of the denominators of two or more fractions. For example, the fractions $\frac{1}{2}$ and $\frac{2}{3}$ have common denominators 6, 12, 18, and other multiples of 6. Fractions with the same denominator already have a common denominator.

common factor A factor of two or more counting numbers. For example, 4 is a common factor of 8 and 12.

discount The amount by which a price of an item is reduced in a sale, usually given as a fraction or percent of the original price, or a percent off. For example, a $4 item on sale for $2 is discounted by 50%, or $\frac{1}{2}$. A $10.00 item at "10% off!" costs $9.00, or $\frac{1}{10}$ less than the usual price.

equivalent fractions Fractions with different denominators that name the same number.

greatest common factor (GCF) The largest factor that two or more counting numbers have in common. For example, the common factors of 24 and 36 are 1, 2, 3, 4, 6, and 12, and their greatest common factor is 12.

improper fraction A fraction whose numerator is greater than or equal to its denominator. For example, $\frac{4}{3}$, $\frac{5}{2}$, $\frac{4}{4}$, and $\frac{24}{12}$ are improper fractions.

In *Everyday Mathematics,* improper fractions are sometimes called top-heavy fractions.

interest A charge for the use of someone else's money. Interest is usually a percentage of the amount borrowed.

least common denominator (LCD) The least common multiple of the denominators of every fraction in a given collection. For example, the least common denominator of $\frac{1}{2}$, $\frac{4}{5}$, and $\frac{3}{8}$ is 40.

least common multiple (LCM) The smallest number that is a multiple of two or more given numbers. For example, common multiples of 6 and 8 include 24, 48, and 72. The least common multiple of 6 and 8 is 24.

mixed number A number that is written using both a whole number and a fraction. For example, $2\frac{1}{4}$ is a mixed number equal to $2 + \frac{1}{4}$.

percent (%) Per hundred, for each hundred, or out of a hundred. $1\% = \frac{1}{100} = 0.01$. For example, *48% of the students in the school are boys* means that out of every 100 students in the school, 48 are boys.

proper fraction A fraction in which the numerator is less than the denominator. A proper fraction is between −1 and 1. For example, $\frac{3}{4}$, $-\frac{2}{5}$, and $\frac{21}{24}$ are proper fractions. Compare to *improper fraction. Everyday Mathematics* does not emphasize these distinctions.

80

quick common denominator (QCD) The product of the denominators of two or more fractions. For example, the quick common denominator of $\frac{3}{4}$ and $\frac{5}{6}$ is 4 * 6, or 24. In general, the quick common denominator of $\frac{a}{b}$ and $\frac{c}{d}$ is b * d. As the name suggests, this is a quick way to get a *common denominator* for a collection of fractions, but it does not necessarily give the *least common denominator*.

simplest form of a fraction A fraction that cannot be renamed in simpler form. Also called

"lowest terms." A mixed number is in simplest form if its fractional part is in simplest form. Simplest form is not emphasized in *Everyday Mathematics* because other equivalent forms are often equally or more useful. For example, when comparing or adding fractions, fractions with a common denominator are likely to be easier to work with than fractions in simplest form.

Do-Anytime Activities

Try these ideas to help your child with the concepts taught in this unit.

1. Consider allowing your sixth grader to accompany you on shopping trips when you know there is a sale. Have him or her bring a calculator to figure out the sale price of items. Ask your child to show you the sale price of the item and the amount of the discount. If your child enjoys this activity, you might extend it by letting him or her calculate the total cost of an item after tax has been added to the subtotal. One way to calculate the total cost is simply to multiply the subtotal by 1.08 (for 8% sales tax). For example, the total cost of a $25 item on which 8% sales tax is levied would be 25 * 1.08 = 25 * (1 + 0.08) = (25 * 1) + (25 * 0.08) = 25 + 2 = 27, or $27.

2. On grocery shopping trips, point out to your child the decimals printed on the item labels on the shelves. These often show unit prices (price per 1 ounce, price per 1 gram, price per 1 pound, and so on), reported to three or four decimal places. Have your child round the numbers to the nearest hundredth (nearest cent).

3. Your child's teacher may display a Fractions, Decimals, Percents Museum in the classroom and expect students to contribute to this exhibit. Help your child look for examples of the ways in which printed advertisements, brochures, and newspaper and magazine articles use fractions, decimals, and percents.

Building Skills through Games

In Unit 4, your child will work on his or her understanding of rational numbers by playing games like the ones described below.

Fraction Action, Fraction Friction See *Student Reference Book*, page 317
Two or three players gather fraction cards that have a sum as close as possible to 2, without going over. Students can make a set of 16 cards by copying fractions onto index cards.

Frac-Tac-Toe See *Student Reference Book*, pages 314–316
Two players need a deck of number cards with 4 each of the numbers 0–10; a game board, a 5-by-5 grid that resembles a bingo card; a *Frac-Tac-Toe* Number-Card board; markers or counters in two different colors, and a calculator. The different versions of *Frac-Tac-Toe* help students practice conversions between fractions, decimals, and percents.

As You Help Your Child with Homework

As your child brings assignments home, you may want to go over the instructions together, clarifying them as necessary. The answers listed below will guide you through some of this unit's Study Links.

Study Link 4◆1

Sample answers for problems 1–16.

1. $\frac{8}{10}$ 2. $\frac{14}{20}$ 3. $\frac{2}{8}$ 4. $\frac{4}{6}$

5. $\frac{10}{8}$ 6. $\frac{4}{4}$ 7. $\frac{3}{4}$ 8. $\frac{1}{5}$

9. $\frac{1}{4}$ 10. $\frac{5}{2}$ 11. $\frac{1}{5}$ 12. $\frac{2}{3}$

13. $\frac{2}{6}, \frac{3}{9}, \frac{4}{12}$ 14. $\frac{3}{4}, \frac{15}{20}, \frac{150}{200}$ 15. $\frac{6}{1}, \frac{12}{2}, \frac{18}{3}$

16. $\frac{24}{10}, \frac{36}{15}, \frac{48}{20}$ 17. $\frac{1}{2}$ 18. $\frac{2}{3}$

19. $\frac{1}{5}$ 20. $\frac{2}{5}$ 21. $\frac{3}{8}$ 22. $\frac{2}{7}$

23. $x = 3$ 24. $y = 12$ 25. $m = 30$

26. $27\frac{1}{4}$ 27. $29\frac{1}{5}$ 28. $29\frac{2}{7}$

Study Link 4◆2

1. > 2. > 3. < 4. <

5. > 6. < 7. $\frac{1}{3}, \frac{2}{5}, \frac{12}{25}$

8. $\frac{1}{12}, \frac{1}{5}, \frac{1}{3}, \frac{2}{5}, \frac{7}{14}, \frac{6}{10}, \frac{15}{16}, \frac{49}{50}$

9. 9.897 10. 3.832 11. 0.823 12. 4.357

Study Link 4◆3

1. $\frac{1}{2}$ 2. $1\frac{1}{16}$ 3. $2\frac{13}{20}$ 4. $\frac{2}{3}$

5. $\frac{11}{12}$ 6. $1\frac{1}{6}$ 7. $1\frac{8}{45}$ 8. 2

9. $\frac{3}{8}$ 10. $1\frac{4}{15}$ 11. $\frac{1}{3}$ 12. $\frac{1}{2}$

13. $1\frac{3}{4}$ 14. $\frac{1}{10}$ 15. 2.7 16. 0.58

17. 1.98

Study Link 4◆4

1. **a.** Sample answer: They may have added only the numerators.

   b. Sample answer: Both fractions are close to 1, so their sum should be close to 2.

2. $1\frac{1}{4}$ inches

3. Sample answer: He can use three $\frac{1}{2}$-cup measures and one $\frac{1}{4}$-cup measure.

4. $4\frac{1}{2}$ 5. $1\frac{3}{4}$ 6. $2\frac{1}{3}$

7. $1\frac{7}{4}, \frac{11}{4}$ 8. 90 9. 246 10. 432

11. 315

Study Link 4◆5

1. **a.** $8\frac{1}{2}$ in. **b.** $1\frac{1}{2}$ in.; $\frac{1}{4}$ in.

2. **a.** $2\frac{1}{2}$ bushels **b.** 30 quarts

3. 4 4. $\frac{2}{3}$ 5. $5\frac{1}{6}$ 6. $\frac{5}{9}$

7. $1\frac{5}{8}$ 8. 6 9. $6\frac{3}{5}$ 10. $1\frac{5}{12}$

11. $2\frac{11}{20}$ 12. 14 13. 17.9 14. $21.99

15. 20

Study Link 4◆6

1. $\frac{6}{20}$ 2. $\frac{15}{63}$ 3. $\frac{15}{8}$ or $1\frac{7}{8}$ 4. $\frac{11}{48}$

5. $\frac{35}{48}$ 6. $\frac{21}{100}$ 7. $\frac{14}{45}$

8. $\frac{32}{7}$ or $4\frac{4}{7}$ 9. $\frac{96}{11}$, or $8\frac{8}{11}$ 10. $\frac{1}{5}$ of the points

11. $2\frac{1}{4}$ cups 12. $\frac{7}{12}$ of the sixth graders

13. **a.** $\frac{1}{2}$ the girls **b.** 6 girls

14. 9 15. 0.1 16. 0.1

Study Link 4·7

1. $\frac{9}{5}$ 2. $\frac{18}{6}$ 3. $\frac{17}{3}$ 4. $\frac{7}{2}$

5. 3 6. $4\frac{1}{8}$ 7. $2\frac{1}{2}$ 8. $6\frac{2}{3}$

9. 3 10. $4\frac{1}{5}$ 11. $2\frac{1}{12}$ 12. $5\frac{4}{9}$

13. $7\frac{31}{32}$ 14. 20 15. 28 16. 63

17. 63

Study Link 4·8

1. $\frac{8}{10}$, 80% 2. $\frac{75}{100}$, 75% 3. $\frac{30}{100}$, $\frac{3}{10}$

4. 0.5 5. 0.75 6. 0.25 7. 1.8

8. $\frac{2}{5}$ 9. $\frac{1}{10}$ 10. $\frac{17}{25}$ 11. $\frac{1}{4}$

12. 50% 13. 25% 14. 60% 15. 95%

16. $\frac{50}{100}$, $\frac{1}{2}$ 17. $\frac{40}{100}$, $\frac{2}{5}$ 18. $\frac{100}{100}$, 1 19. $\frac{180}{100}$, $1\frac{4}{5}$

Study Link 4·9

1. 65% 2. 33.4% 3. 2% 4. 40%

5. 270% 6. 309% 7. 0.27 8. 0.539

9. 0.08 10. 0.60 11. 1.80 12. 1.15

13. 0.88, 88% 14. 0.42, 42%

Study Link 4·10

Problems 1–4 are circle graphs.

Study Link 4·11

1. Table entries: 150, 100, 125, 125 students

2. Table entries: 18, 12, 15, 15 students

3. a. 3.3 b. 8.8 c. 22

Equivalent Fractions

Find an equivalent fraction by multiplying.

1. $\frac{4}{5}$ _____

2. $\frac{7}{10}$ _____

3. $\frac{1}{4}$ _____

4. $\frac{2}{3}$ _____

5. $\frac{5}{4}$ _____

6. $\frac{2}{2}$ _____

Find an equivalent fraction by dividing.

7. $\frac{9}{12}$ _____

8. $\frac{20}{100}$ _____

9. $\frac{4}{16}$ _____

10. $\frac{30}{12}$ _____

11. $\frac{10}{50}$ _____

12. $\frac{16}{24}$ _____

Write 3 equivalent fractions for each number.

13. $\frac{1}{3}$ _____

14. $\frac{75}{100}$ _____

15. 6 _____

16. $\frac{12}{5}$ _____

Write each fraction in simplest form.

17. $\frac{8}{16}$ _____

18. $\frac{6}{9}$ _____

19. $\frac{3}{15}$ _____

20. $\frac{10}{25}$ _____

21. $\frac{6}{16}$ _____

22. $\frac{14}{49}$ _____

Find the missing numbers.

23. $\frac{1}{5} = \frac{x}{15}$

24. $\frac{2}{3} = \frac{y}{18}$

25. $\frac{15}{25} = \frac{m}{50}$

$x =$ _____

$y =$ _____

$m =$ _____

Practice

Divide. Express the remainder as a fraction in simplest form.

26. $24\overline{)654}$

27. $25\overline{)730}$

28. $14\overline{)410}$

Comparing and Ordering Fractions

Write <, >, or = to make a true number sentence. For each problem that you did not solve mentally, show how you got the answer.

1. $\frac{4}{5}$ _____ $\frac{2}{5}$

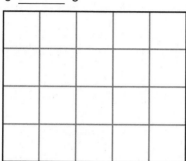

2. $\frac{3}{8}$ _____ $\frac{1}{3}$

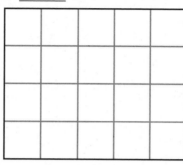

3. $\frac{3}{4}$ _____ $\frac{17}{20}$

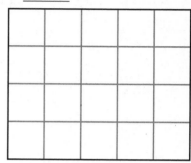

4. $\frac{19}{20}$ _____ $\frac{99}{100}$

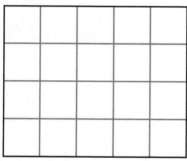

5. $\frac{4}{7}$ _____ $\frac{4}{10}$

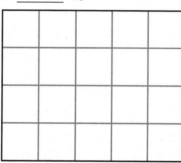

6. $\frac{2}{3}$ _____ $\frac{7}{9}$

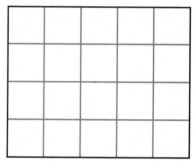

7. Circle each fraction that is less than $\frac{1}{2}$. $\quad \frac{3}{6} \quad \frac{6}{10} \quad \frac{1}{3} \quad \frac{2}{5} \quad \frac{6}{11} \quad \frac{12}{25}$

8. Write the fractions in order from smallest to largest.

$$\frac{1}{3} \quad \frac{15}{16} \quad \frac{7}{14} \quad \frac{1}{5} \quad \frac{2}{5} \quad \frac{49}{50} \quad \frac{6}{10} \quad \frac{1}{12}$$

_____ _____ _____ _____ _____ _____ _____ _____

Practice

9. $13.987 - 4.09 = $ _____

10. $5.9 - 2.068 = $ _____

11. $0.9 - 0.077 = $ _____

12. $8 - 3.643 = $ _____

STUDY LINK 4·3 Adding and Subtracting Fractions

Add or subtract. Write each answer in simplest form. If possible, rename answers as mixed numbers or whole numbers.

1. $\frac{1}{3} + \frac{1}{6} =$ _____

2. $\frac{3}{4} + \frac{5}{16} =$ _____

3. $\frac{9}{4} + \frac{2}{5} =$ _____

4. $\frac{2}{9} + \frac{4}{9} =$ _____

5. $\frac{1}{6} + \frac{3}{4} =$ _____

6. $\frac{5}{12} + \frac{3}{4} =$ _____

7. $\frac{7}{9} + \frac{2}{5} =$ _____

8. $\frac{5}{4} + \frac{3}{4} =$ _____

9. $\frac{7}{8} - \frac{2}{4} =$ _____

10. $\frac{5}{3} - \frac{2}{5} =$ _____

11. $\frac{11}{12} - \frac{7}{12} =$ _____

12. $\frac{4}{5} - \frac{3}{10} =$ _____

13. $\frac{15}{8} - \frac{3}{24} =$ _____

14. $\frac{3}{5} - \frac{1}{2} =$ _____

Practice

Solve mentally.

15. $3 - 0.30 =$ _____

16. $0.60 - 0.02 =$ _____

17. $2 - 0.02 =$ _____

89

Name   Date   Time

STUDY LINK 4·4   +, − Fractions and Mixed Numbers

1. In a national test, eighth-grade students answered the problem shown in the top of the table at the right. Also shown are the 5 possible answers they were given and the percent of students who chose each answer.

   a. What mistake do you think the students who chose C made?

   _____

   _____

   b. Explain why B is the best estimate.

   _____

   _____

   _____

	Estimate the answer to $\frac{12}{13} + \frac{7}{8}$. You will not have enough time to solve the problem using paper and pencil.
Possible Answers	**Percent Who Chose This Answer**
A. 1	7%
B. 2	24%
C. 19	28%
D. 21	27%
E. I don't know.	14%

2. A board is $6\frac{3}{8}$ inches long. Verna wants to cut enough so that it will be $5\frac{1}{8}$ inches long. How much should she cut? _____ (unit)

3. Tim is making papier-mâché. The recipe calls for $1\frac{3}{4}$ cups of paste. Using only $\frac{1}{2}$-cup, $\frac{1}{4}$-cup, and $\frac{1}{3}$-cup measures, how can he measure the correct amount?

_____

_____

Add or subtract. Write your answers as mixed numbers in simplest form. Show your work on the back of the page. Use number sense to check whether each answer is reasonable.

4. $3\frac{1}{4} + 1\frac{1}{4} =$ _____

5. $4 - 2\frac{1}{4} =$ _____

6. $1\frac{2}{3} + \frac{2}{3} =$ _____

7. Circle the numbers that are equivalent to $2\frac{3}{4}$.

$1\frac{7}{4}$ $\frac{6}{4}$ $\frac{3}{7}$ $\frac{11}{4}$

Practice

Solve mentally.

8. $5 * 18 =$ _____

9. $6 * 41 =$ _____

10. $9 * 48 =$ _____

11. $7 * 45 =$ _____

91

Copyright © Wright Group/McGraw-Hill

STUDY LINK 4·5 Mixed-Number Practice

SRB 84–86

1. Answer the following questions about the rectangle shown at the right. Include units in your answers.

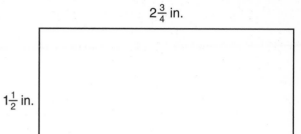

$2\frac{3}{4}$ in.

$1\frac{1}{2}$ in.

a. What is the perimeter? _____

b. If you were to trim this rectangle so that it was a square measuring $1\frac{1}{4}$ inches on a side, how much would you cut

from the base? _____ from the height? _____

2. Michael bought 1 peck of Empire apples, 1 peck of Golden Delicious apples, a $\frac{1}{2}$-bushel of Red Delicious apples, and $1\frac{1}{2}$ bushels of McIntosh apples.

$1 \text{ peck} = \frac{1}{4} \text{ bushel}$

a. How many bushels of apples did he buy in all? _____

b. Michael estimates that he can make about 12 quarts of applesauce per bushel of apples. About how many quarts of applesauce can he make from the apples he bought? _____

Add or subtract. Show your work and estimates on the back of the page.

3. $2\frac{1}{3} + 1\frac{2}{3} =$ _____

4. $6\frac{1}{3} - 5\frac{2}{3} =$ _____

5. $4\frac{1}{2} + \frac{2}{3} =$ _____

6. $6 - 5\frac{4}{9} =$ _____

7. $4\frac{3}{8} - 2\frac{3}{4} =$ _____

8. $3\frac{1}{4} + 2\frac{3}{4} =$ _____

9. $9 - 2\frac{2}{5} =$ _____

10. $4\frac{1}{4} - 2\frac{5}{6} =$ _____

11. $5\frac{1}{4} - 2\frac{7}{10} =$ _____

Practice

Solve mentally.

12. $1\frac{1}{2} + 4\frac{2}{3} + 2\frac{1}{2} + 5\frac{1}{3} =$ _____

13. $4.5 + 3.4 + 7.5 + 2.5 =$ _____

14. $\$2.35 + \$9.60 + \$8.05 + \$1.99 =$ _____

15. $5\frac{5}{8} + 3\frac{3}{4} + 2\frac{1}{4} + 8\frac{3}{8} =$ _____

93

STUDY LINK 4·6 Fraction Multiplication

Use the fraction multiplication algorithm below to solve the following problems.

> **Fraction Multiplication Algorithm**
>
> $$\frac{a}{b} * \frac{c}{d} = \frac{a * c}{b * d}$$

1. $\frac{3}{5} * \frac{2}{4} =$ _____

2. $\frac{3}{7} * \frac{5}{9} =$ _____

3. $5 * \frac{3}{8} =$ _____

4. _____ $= \frac{11}{12} * \frac{1}{4}$

5. $\frac{5}{6} * \frac{7}{8} =$ _____

6. $\frac{3}{10} * \frac{7}{10} =$ _____

7. _____ $= \frac{2}{5} * \frac{7}{9}$

8. $\frac{4}{7} * 8 =$ _____

9. $12 * \frac{8}{11} =$ _____

10. South High beat North High in basketball, scoring $\frac{4}{5}$ of the total points. Rachel scored $\frac{1}{4}$ of South High's points. What fraction of the total points did Rachel score?

_____

11. Josh was making raisin muffins for a party. He needed to triple the recipe, which called for $\frac{3}{4}$ cup raisins. How many cups of raisins did he need?

_____

12. At Long Middle School, $\frac{7}{8}$ of the sixth graders live within 1 mile of the school. About $\frac{2}{3}$ of those sixth graders walk to school. None who live a mile or more away walk to school. About what fraction of the sixth graders walk to school?

_____

13. a. For Calista's 12th birthday party, her mom will order pizza. $\frac{3}{4}$ of the girls invited like vegetables on their pizza. However, $\frac{1}{3}$ of those girls won't eat green peppers. What fraction of all the girls will eat a green-pepper-and-onion pizza?

_____

b. If 12 girls are at the party (including Calista), how many girls will not eat a green-pepper-and-onion pizza?

_____

Practice

Solve.

14. $12 * 0.75 =$ _____

15. $0.2 * 0.5 =$ _____

16. $0.4 * 0.25 =$ _____

STUDY LINK
4·7

Multiplying Mixed Numbers

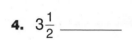

Rename each mixed number as a fraction.

1. $1\frac{4}{5}$ _____

2. $2\frac{6}{6}$ _____

3. $5\frac{2}{3}$ _____

4. $3\frac{1}{2}$ _____

Rename each fraction as a mixed number or whole number.

5. $\frac{12}{4}$ _____

6. $\frac{33}{8}$ _____

7. $\frac{15}{6}$ _____

8. $\frac{20}{3}$ _____

Multiply. Write each answer in simplest form. If possible, write answers as mixed numbers or whole numbers.

9. $5 * \frac{3}{5} =$ _____

10. $2\frac{1}{3} * 1\frac{4}{5} =$ _____

11. $\frac{5}{6} * 2\frac{1}{2} =$ _____

12. $1\frac{1}{6} * 4\frac{2}{3} =$ _____

13. $3\frac{3}{4} * 2\frac{1}{8} =$ _____

14. $7\frac{1}{2} * 2\frac{2}{3} =$ _____

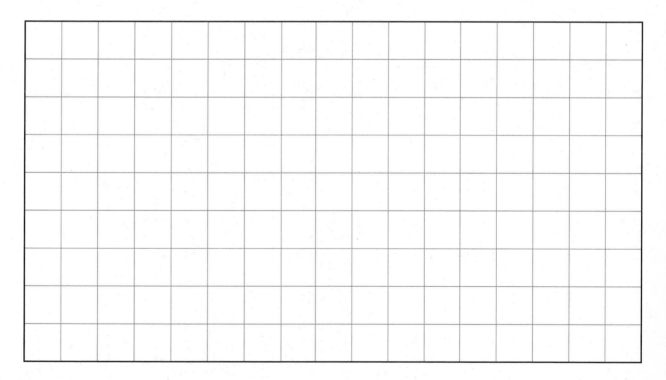

Practice

Solve mentally.

15. $8 * 3.5 =$ _____

16. $12 * 5.25 =$ _____

17. $4.2 * 15 =$ _____

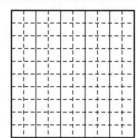

Fractions, Decimals, and Percents

Fill in the missing numbers below. Then shade each large square to represent all three of the equivalent numbers below it. Each large square is worth 1.

1.

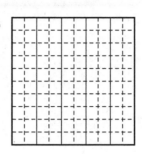

$$\frac{4}{5} = \frac{\boxed{}}{10} = \underline{\hspace{1cm}}\%$$

2.

$$\frac{6}{8} = \frac{\boxed{}}{100} = \underline{\hspace{1cm}}\%$$

3.

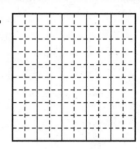

$$30\% = \frac{\boxed{}}{100} = \frac{\boxed{}}{10}$$

Rename the fractions as decimals.

4. $\frac{7}{14} = $ _____

5. $\frac{6}{8} = $ _____

6. $\frac{5}{20} = $ _____

7. $1\frac{4}{5} = $ _____

Rename the decimals as fractions in simplest form.

8. $0.4 = $ _____

9. $0.10 = $ _____

10. $0.68 = $ _____

11. $0.25 = $ _____

Rename the fractions as percents.

12. $\frac{25}{50} = $ _____

13. $\frac{6}{24} = $ _____

14. $\frac{18}{30} = $ _____

15. $\frac{19}{20} = $ _____

Rename the percents as fractions in simplest form.

16. $50\% = \frac{\boxed{}}{100} = $ _____

17. $40\% = \frac{\boxed{}}{100} = $ _____

18. $100\% = \frac{\boxed{}}{100} = $ _____

19. $180\% = \frac{\boxed{}}{100} = $ _____

Experiment

People often don't realize that fractions, decimals, and percents are numbers. To them, numbers are whole numbers like 1, 5, or 100. Try the following experiment: Ask several adults to name four numbers between 1 and 10. Then ask several children. Keep a record of all responses on the back of this page. How many named fractions, decimals, or percents? Now ask the same people to name four numbers between 1 and 3. Report your findings.

STUDY LINK 4·9 Decimals, Percents, and Fractions

Rename each decimal as a percent.

1. 0.65 = _____

2. 0.334 = _____

3. 0.02 = _____

4. 0.4 = _____

5. 2.7 = _____

6. 3.09 = _____

Rename each percent as a decimal.

7. 27% = _____

8. 53.9% = _____

9. 8% = _____

10. 60% = _____

11. 180% = _____

12. 115% = _____

Use division to rename each fraction as a decimal to the nearest hundredth.
Then rename the decimal as a percent.

13. $\frac{7}{8}$ = 0._____ = _____%

14. $\frac{5}{12}$ = 0._____ = _____%

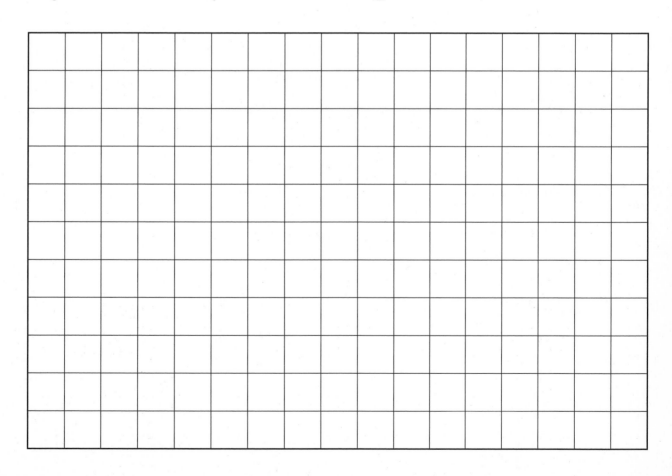

STUDY LINK 4·10 | Circle Graphs

Use estimation to make a circle graph displaying the data in each problem. (*Hint:* For each percent, think of a simple fraction that is close to the value of the percent. Then estimate the size of the sector for each percent.) Remember to graph the smallest sector first.

1. According to the 2000 Census, 21.2% of the U.S. population was under the age of 15, 12.6% was age 65 or older, and 66.2% was between the ages of 15 and 64.

Age of U.S. Population

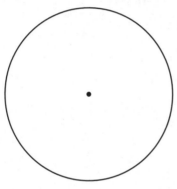

2. In 2004, NASA's total budget was $15.4 billion. 51% was spent on Science, Aeronautics, and Exploration. 48.8% was spent on Space Flight Capabilities, and 0.2% was spent on the Inspector General.

NASA Budget

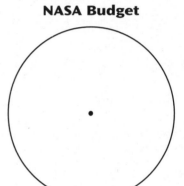

3. 98.3% of households in the United States have at least one television.

Households with TV

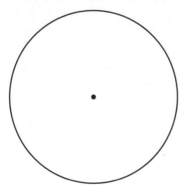

4. The projected school enrollment for the United States in 2009 is 72 million students. 23.2% will be in college, 22.9% will be in high school, and 53.9% will be in Grades Pre-K–8.

U.S. School Enrollment

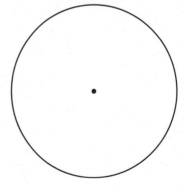

Percent Problems

STUDY LINK 4·11

The results of a survey about children's weekly allowances are shown at the right.

Amount of Allowance	Percent of Children
$0	30%
$1–$4	20%
$5	25%
$6 or more	25%

1. Lincoln School has about 500 students. Use the survey results to complete this table.

Amount of Allowance	Predicted Number of Students at Lincoln
$0	
$1–$4	
$5	
$6 or more	

2. The sixth grade at Lincoln has about 60 students. Use the survey results to complete this table.

Amount of Allowance	Predicted Number of Sixth-Grade Students at Lincoln
$0	
$1–$4	
$5	
$6 or more	

A rule of thumb for changing a number of meters to yards is to add the number of meters to 10% of the number of meters.

Examples: 5 m is about 5 + (10% of 5), or 5.5, yd.

10 m is about 10 + (10% of 10), or 11, yd.

3. Use this rule of thumb to estimate how many yards are in the following numbers of meters.

a. 3 m is about 3 + (10% of 3), or _____, yd.

b. 8 m is about 8 + (10% of 8), or _____, yd.

c. 20 m is about 20 + (10% of 20), or _____, yd.

105

STUDY LINK 4·12 Unit 5: Family Letter

Geometry: Congruence, Constructions, and Parallel Lines

In *Fourth* and *Fifth Grade Everyday Mathematics,* students used a compass and straightedge to construct basic shapes and create geometric designs. In Unit 5 of *Sixth Grade Everyday Mathematics,* students will review some basic construction techniques and then devise their own methods for copying triangles and quadrilaterals and for constructing parallelograms. The term *congruent* will be applied to their copies of line segments, angles, and 2-dimensional figures. Two figures are congruent if they have the *same size* and the *same shape.*

Another approach to congruent figures in Unit 5 is through isometry transformations. These are rigid motions that take a figure from one place to another while preserving its size and shape. Reflections (flips), translations (slides), and rotations (turns) are basic isometry transformations (also known as rigid motions). A figure produced by an isometry transformation (the image) is congruent to the original figure (the preimage).

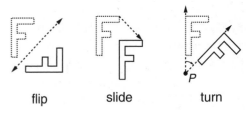

flip   slide   turn

Students will continue to work with the Geometry Template, a tool that was introduced in *Fifth Grade Everyday Mathematics.* The Geometry Template contains protractors and rulers for measuring and cutouts for drawing geometric figures. Students will review how to measure and draw angles using the full-circle and half-circle protractors.

Students will also use a protractor to construct circle graphs that represent data collections. This involves converting the data to percents of a total, finding the corresponding degree measures around a circle, and drawing sectors of the appropriate size.

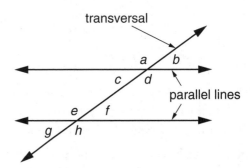

If the measure of any one angle is given, the measures of all the others can be found without measuring.

Measures often can be determined without use of a measuring tool. Students will apply properties of angles and sums of angles to find unknown measures in figures similar to those at the right.

One lesson in Unit 5 is a review and extension of work with the coordinate grid. Students will plot and name points on a 4-quadrant coordinate grid and use the grid for further study of geometric shapes.

Please keep this Family Letter for reference as your child works through Unit 5.

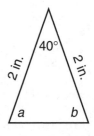

The sum of the angles in a triangle is 180°. Angles *a* and *b* have the same measure, 70°.

107

Math Tools

Your child will use a compass and a straightedge to construct geometric figures. A compass is used to draw a circle, or part of a circle, called an arc. A straightedge is used only to draw straight lines, not for measuring. The primary difference between a compass-and-straightedge construction and a drawing or sketch of a geometric figure is that measuring is not allowed in constructions.

Vocabulary

Important terms in Unit 5:

adjacent angles Two angles with a common side and vertex that do not otherwise overlap. In the diagram, angles *a* and *b* are adjacent angles. So are angles *b* and *c, d* and *a,* and *c* and *d.*

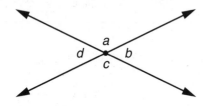

congruent Figures that have exactly the same size and shape are said to be congruent to each other. The symbol ≅ means "is congruent to."

line of reflection (mirror line) A line halfway between a figure (preimage) and its reflected image. In a reflection, a figure is flipped over the line of reflection.

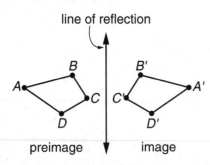

line of reflection

preimage   image

ordered pair Two numbers, or coordinates, used to locate a point on a rectangular coordinate grid. The first coordinate *x* gives the position along the horizontal axis of the grid, and the second coordinate *y* gives the position along the vertical axis. The pair is written (*x,y*).

reflection (flip) The flipping of a figure over a line (line of reflection) so its image is the mirror image of the original (preimage).

reflex angle An angle measuring between 180° and 360°.

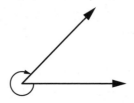

rotation (turn) A movement of a figure around a fixed point or an axis; a turn.

supplementary angles Two angles whose measures add to 180°. Supplementary angles do not need to be adjacent.

translation (slide) A transformation in which every point in the image of a figure is at the same distance in the same direction from its corresponding point in the figure. Informally called a slide.

vertical (opposite) angles The angles made by intersecting lines that do not share a common side. Same as opposite angles. Vertical angles have equal measures. In the diagram, angles 1 and 3 are vertical angles. They have no sides in common. Similarly, angles 4 and 2 are vertical angles.

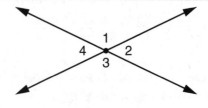

Do-Anytime Activities

To work with your child on the concepts taught in this unit, try these interesting and engaging activities:

1. While you are driving in the car together, ask your child to look for congruent figures, for example, windows in office buildings, circles on stoplights, or wheels on cars and trucks.

2. Look for apparent right angles or any other type of angles: acute (less than 90°) or obtuse (between 90° and 180°). Guide your child to look particularly at bridge supports to find a variety of angles.

3. Triangulation lends strength to furniture. Encourage your child to find corner triangular braces in furniture throughout your home. Look under tables, under chairs, inside cabinets, or under bed frames. Have your child count how many examples of triangulation he or she can find in your home.

Building Skills through Games

In Unit 5, students will work on their understanding of geometry concepts by playing games such as those described below.

Angle Tangle See *Student Reference Book,* page 306
Two players need a protractor, straightedge, and blank paper to play *Angle Tangle*. Skills practiced include estimating angle measures as well as measuring angles.

Polygon Capture See *Student Reference Book*, page 330
Players capture polygons that match both the angle property and the side property drawn. Properties include measures of angles, lengths of sides, and number of pairs of parallel sides.

Students will review concepts from previous units by playing games such as:

2-4-8 and 3-6-9 *Frac-Tac-Toe* (Decimal Versions) See *Student Reference Book*, pages 314–316
Two players need a deck of number cards with four each of the numbers 0–10; a gameboard; a 5 × 5 grid that resembles a bingo card; a *Frac-Tac-Toe* Number-Card board; markers or counters in two different colors; and a calculator. The two versions, *2-4-8 Frac-Tac-Toe* and *3-6-9 Frac-Tac-Toe*, help students practice conversions between fractions and decimals.

As You Help Your Child with Homework

As your child brings assignments home, you may want to go over the instructions together, clarifying them as necessary. The answers listed below will guide you through some of the Unit 5 Study Links.

Study Link 5·1

2. a. ∠H   **b.** ∠J

   c. ∠D   **d.** ∠ABC, ∠GFE, ∠L

3b. 180°   **3c.** 360°

Study Link 5·2

1. m∠y = 120°   **2.** m∠x = 115°

3. m∠c = 135°   m∠a = 45°   m∠t = 135°

4. m∠q = 120°   m∠r = 80°   m∠s = 70°

5. m∠a = 120°   m∠b = 60°   m∠c = 120°

   m∠d = 40°   m∠e = 140°   m∠f = 140°

   m∠g = 80°   m∠h = 100°   m∠i = 100°

6. m∠w = 90°   m∠a = 75°   m∠t = 105°

   m∠c = 75°   m∠h = 105°

7. 12   **8.** 30   **9.** 110

Study Link 5·3

2. a. 1,920,000 adults   **b.** 3,760,000 adults

3. −7, 0, 0.07, 0.7, 7

4. 0.06, $\frac{1}{10}$, 0.18, 0.2, 0.25, 0.75, $\frac{4}{5}$, $\frac{4}{4}$

Study Link 5·4

Sample answers for 1–3:

1. Vertex C: (1,2)

2. Vertex F: (5,10)   Vertex G: (3,7)

3. Vertex J: (2,1)   **4.** Vertex M: (−2,−3)

5. Vertex Q: (8,−3)

Study Link 5·5

1.   **2.**   **3.**

4. 64   **5.** 243   **6.** 1   **7.** 64

Study Link 5·7

1. m∠r = 47°   m∠s = 133°   m∠t = 47°

2. m∠NKO = 10°

3. m∠a = 120°   m∠b = 120°   m∠c = 60°

4. m∠a = 57°   m∠c = 114°   m∠t = 57°

5. m∠x = 45°   m∠y = 45°   m∠z = 135°

6. m∠p = 54°

7. 0.0027   **8.** 0.12   **9.** 0.0049   **10.** 0.225

Study Link 5·8

2. A': (−2,−7)   B': (−6,−6)

   C': (−8,−4)   D': (−5,−1)

3. A": (2,1)   B": (6,2)

   C": (8,4)   D": (5,7)

4. A'": (1,−2)   B'": (2,−6)

   C'": (4,−8)   D'": (7,−5)

5. 0.3   **6.** 0.143   **7.** 0.0359

Study Link 5·9

3. Sample answers: All of the vertical angles have the same measure; all of the angles along the transversal and on the same side are supplementary; opposite angles along the transversal are equal in measure.

Study Link 5·10

1. a. 50°; ∠YZW plus the 130° angle equals 180°, so ∠YZW = 50°. Because opposite angles in a parallelogram are equal, ∠X also equals 50°.

   b. 130°; m∠YZW = 50° and ∠Y and ∠Z are consecutive angles. Because consecutive angles of parallelograms are supplementary, ∠Y = 130°.

2. Opposite sides of a parallelogram are congruent.

3. 110°; Adjacent angles that form a straight angle are supplementary.

4. square   **5.** rhombus

STUDY LINK
5·1 **Angles**

1. Measure each angle to the nearest degree. Write the measure next to the angle.

SRB
160
230–232

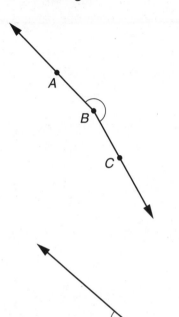

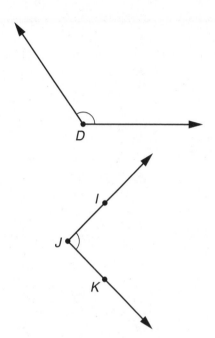

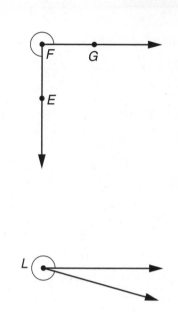

2. **a.** Which angle above is an acute angle? _____

   b. A right angle? _____   **c.** An obtuse angle? _____

   d. Which angles above are reflex angles? _____

3. **a.** Measure each angle in triangle *ADB* at the right.

   b. Find the sum of the 3 angle measures. _____

   c. Use Problem 3b to calculate the sum of the
   interior angle measures in quadrangle *ABCD*. _____

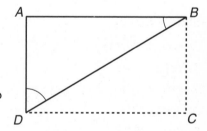

Try This

4. Find the measure of ∠*KLM*. Then draw an angle that is 60% of the
   measure of ∠*KLM* on the reverse side of this paper. Label it as ∠*NOP*.

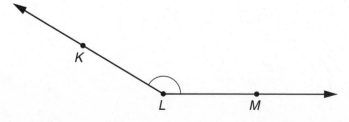

111

STUDY LINK 5·2 Angle Relationships

Find the following angle measures. Do not use a protractor.

SRB
163 233

1.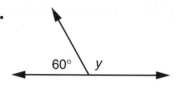

60° y

m∠y = _____

2.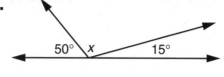

50° x   15°

m∠x = _____

3.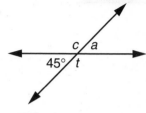

c a
45° t

m∠c = _____

m∠a = _____

m∠t = _____

4. m∠q = _____   m∠r = _____   m∠s = _____

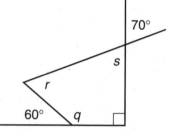

70°
s
r
60° q

5. m∠a = _____   m∠b = _____

m∠c = _____   m∠d = _____

m∠e = _____   m∠f = _____

m∠g = _____   m∠h = _____

m∠i = _____

g
h i
80°
c b   40°   f
60° a   e   d

6. m∠w = _____   m∠a = _____

m∠t = _____   m∠c = _____

m∠h = _____

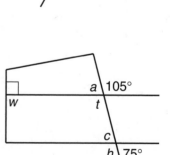

a 105°
w   t
c
h 75°

Practice

7. $\frac{3}{4}$ of 16 = _____

8. $\frac{3}{5}$ of 50 = _____

9. $\frac{1}{3}$ of 330 = _____

STUDY LINK 5·3 | Circle Graphs

1. The table below shows a breakdown, by age group, of adults who listen to classical music.

   a. Calculate the degree measure of each sector to the nearest degree.

   b. Use a protractor to make a circle graph. Do *not* use the Percent Circle. Write a title for the graph.

Age	Percent of Listeners	Degree Measure
18–24	11%	
25–34	18%	
35–44	24%	
45–54	20%	
55–64	11%	
65+	16%	

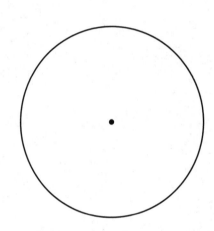

Source: USA Today, Snapshot

2. On average, about 8 million adults listen to classical music on the radio each day.

   a. Estimate how many adults between the ages of 35 and 44 listen to classical music on the radio each day.

   About _____
   (unit)

   b. Estimate how many adults at least 45 years old listen to classical music on the radio each day.

   About _____
   (unit)

Practice

Order each set of numbers from least to greatest.

3. 7, 0.07, −7, 0.7, 0 _____

4. 0.25, 0.75, 0.2, $\frac{4}{5}$, $\frac{4}{4}$, 0.06, 0.18, $\frac{1}{10}$

STUDY LINK 5·4

More Polygons on a Coordinate Grid

SRB
166 169
234

For each polygon described below, some vertices are plotted on the grid.
Either one vertex or two vertices are missing.

◆ Plot and label the missing vertex or vertices on the grid.
  (There may be more than one place you can plot a point.)

◆ Write an ordered number pair for each vertex you plot.

◆ Draw the polygon.

1. Right triangle ABC Vertex C: (_____,_____)

2. Parallelogram $DEFG$ Vertex F: (_____,_____)   Vertex G: (_____,_____)

3. Scalene triangle HIJ Vertex J: (_____,_____)

4. Kite $KLMN$ Vertex M: (_____,_____)   **5.** Square $PQRS$ Vertex Q: (_____,_____)

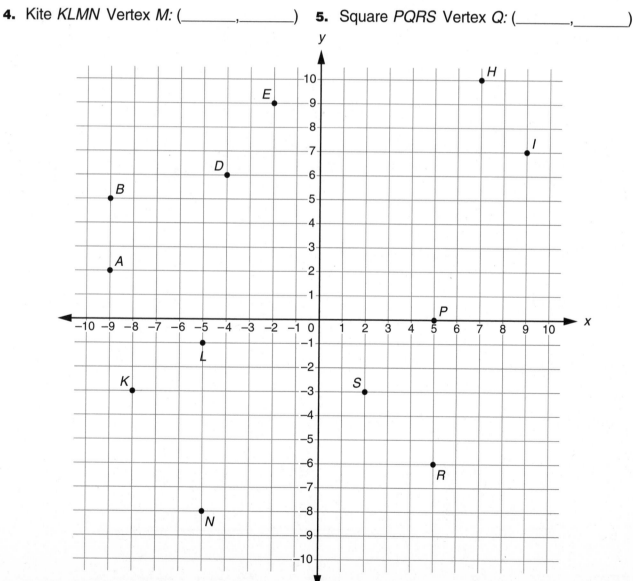

117

STUDY LINK
5·5
Transforming Patterns

A pattern can be translated, reflected, or rotated to create many different designs. Consider the pattern at the right.

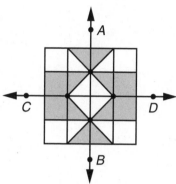

The following examples show how the pattern can be transformed to create different designs:

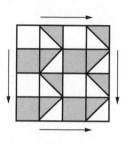

Translations

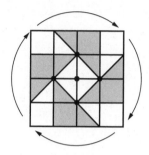

Rotations

Reflections

1. Translate the pattern at the right across 2 grid squares. Then translate the resulting pattern (the given pattern and its translation) down 2 grid squares.

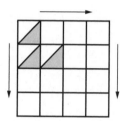

2. Rotate the given pattern clockwise 90° around point *X*. Repeat 2 more times.

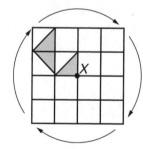

3. Reflect the given pattern over line *JK*. Reflect the resulting pattern (the given pattern and its reflection) over line *LM*.

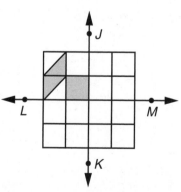

Practice

4. $2^6 =$ _____

5. $3^5 =$ _____

6. $7^0 =$ _____

7. $4^3 =$ _____

119

Congruent Figures and Copying

Column 1 below shows paths with the Start points marked. Complete each path in Column 2 so that it is congruent to the path in Column 1. Use the Start points marked in Column 2. In Problems 2 and 3, the copy will not be in the same position as the original path.

(*Hint:* If you have trouble, try tracing the path in Column 1 and then slide, flip, or rotate it so that its starting point matches the starting point in Column 2.)

Example: These two paths are congruent, but they are not in the same position.

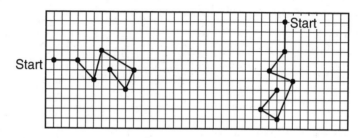

Column 1   Column 2

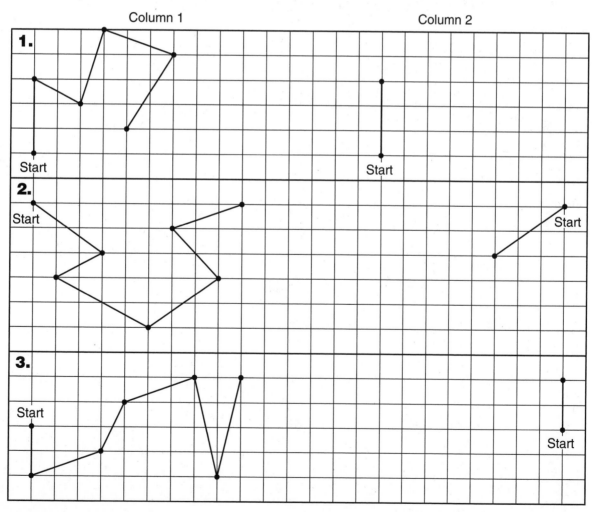

121

STUDY LINK 5·7   **Angle Relationships**

Write the measures of the angles indicated in Problems 1–6.
Do not use a protractor.

1.

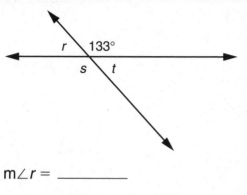

m∠r = _____

m∠s = _____

m∠t = _____

2. ∠*JKL* is a straight angle.

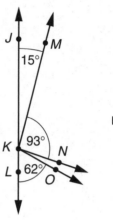

m∠*NKO* = _____

3.

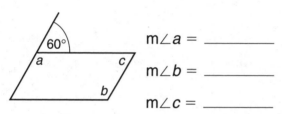

m∠a = _____

m∠b = _____

m∠c = _____

4. Angles *a* and *t* have
the same measure.

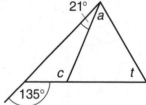

m∠a = _____

m∠c = _____

m∠t = _____

5. Angles *x* and *y* have the
same measure.

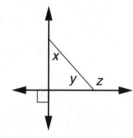

m∠x = _____

m∠y = _____

m∠z = _____

6.

m∠p = _____

Practice

7. 0.09 * 0.03 = _____

8. 0.15 * 0.8 = _____

9. 0.07 * 0.07 = _____

10. 0.75 * 0.3 = _____

STUDY LINK 5·8 **Isometry Transformations on a Grid**

1. Graph and label the following points on the coordinate grid.
   Connect the points to form quadrangle *ABCD*.

   A: (−2,1)   *B:* (−6,2)
   C: (−8,4)   *D:* (−5,7)

2. Translate each vertex of
   ABCD (in Problem 1)
   0 units to the left or right
   and 8 units down. Plot
   and connect the new
   points. Label them
   A′, *B′*, *C′*, and *D′*.

   Record the coordinates of
   the image.

   _____

   _____

   _____

   _____

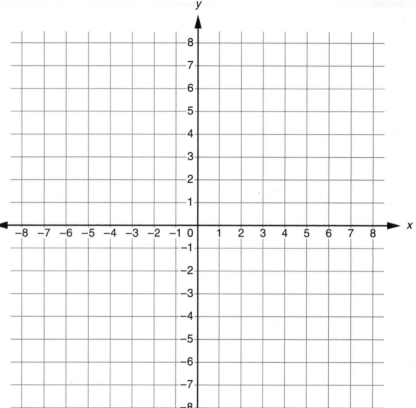

3. Reflect quadrangle *ABCD* across the *y*-axis. Plot and connect the new points.
   Label them *A″*, *B″*, *C″*, and *D″*. Record the coordinates of the image.

   _____   _____   _____   _____

Try This

4. Rotate quadrangle *A″B″C″D″* 90° clockwise around point (0,0). Plot and connect the new
   points. Label them *A‴*, *B‴*, *C‴*, and *D‴*. Record the coordinates of the rotated image.

   _____   _____   _____   _____

Practice

5. $300 * 0.001 =$ _____

6. $143 * 10^{-3} =$ _____

7. $35.9 * \frac{1}{1,000} =$ _____

STUDY LINK 5·9 | **Parallel Lines and a Transversal**

1. Use a ruler and a straightedge to draw 2 parallel lines. Then draw another line that crosses both parallel lines.

2. Measure the 8 angles in your figure.
   Write each measure inside the angle.

3. What patterns do you notice in your angle measures?

   _____

   _____

   _____

   _____

   _____

   _____

Practice

Remember: 1,000 milliliters (mL) = 1 liter (L)

4. 500 mL = _____ L

5. 2.5 L = _____ mL

6. 1,300 mL = _____ L

7. 0.95 L = _____ mL

8. 3,250 mL = _____ L

9. 0.045 L = _____ mL

127

STUDY LINK 5·10 | Parallelogram Problems

All of the figures on this page are parallelograms.
Do not use a ruler or a protractor to solve Problems 1, 2, or 3.

1. a. The measure of ∠X = _____°. Explain how you know.

_____

_____

_____

_____

b. The measure of ∠Y = _____°. Explain how you know.

_____

_____

_____

2. Alexi said that the only way to find the length of sides *CO* and *OA* is to measure them with a ruler. Explain why he is incorrect.

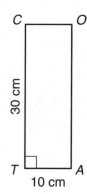

_____

_____

3. What is the measure of ∠*MAR*? _____. Explain how you know.

_____

_____

4. Draw a parallelogram in which all sides have the same length and all angles have the same measure.

What is another name for this parallelogram? _____

5. Draw a parallelogram in which all sides have the same length and no angle measures 90°.

What is another name for this parallelogram? _____

STUDY LINK 5·11 | **Unit 6: Family Letter**

Number Systems and Algebra Concepts

In *Fourth* and *Fifth Grade Everyday Mathematics,* your child worked with addition and subtraction of positive and negative numbers. In this unit, students use multiplication patterns to help them establish the rules for multiplying and dividing with positive and negative numbers. They also develop and use an algorithm for the division of fractions.

In the rest of the unit, your child will explore beginning algebra concepts. First, the class reviews how to determine whether a number sentence is true or false. This involves understanding what to do with numbers that are grouped within parentheses and knowing in what order to calculate if the groupings of numbers are not made explicit by parentheses.

Students then solve simple equations by trial and error to reinforce what it means to solve an equation—to replace a variable with a number that will make the number sentence true.

Next, they solve pan-balance problems, first introduced in *Fifth Grade Everyday Mathematics,* to develop a more systematic approach to solving equations. For example, to find out how many marbles weigh as much as 1 orange in the top balance at the right, you can first remove 1 orange from each pan and then remove half the remaining oranges from the left side and half the marbles from the right side. The pans will still balance.

Students learn that each step in the solution of a pan-balance problem can be represented by an equation, thus leading to the solution of the original equation. You might ask your child to demonstrate how pan-balance problems work.

Finally, your child will learn how to solve inequalities—number sentences comparing two quantities that are not equal.

Please keep this Family Letter for reference as your child works through Unit 6.

Vocabulary

Important terms in Unit 6:

cover-up method An informal method for finding the solution of an open sentence by covering up a part of the sentence containing a variable.

Division of Fractions Property A property of dividing that says division by a fraction is the same as multiplication by the *reciprocal* of the fraction. Another name for this property is the "invert and multiply rule." For example:

$$5 \div 8 = 5 * \frac{1}{8} = \frac{5}{8}$$
$$15 \div \frac{3}{5} = 15 * \frac{5}{3} = \frac{75}{3} = 25$$
$$\frac{1}{2} \div \frac{3}{5} = \frac{1}{2} * \frac{5}{3} = \frac{5}{6}$$

In symbols: For *a* and nonzero *b*, *c*, and *d*,

$$\frac{a}{b} \div \frac{c}{d} = \frac{a}{b} * \frac{d}{c}$$

If $b = 1$, then $\frac{a}{b} = a$ and the property is applied as in the first two examples above.

equivalent equations Equations with the same solution. For example, $2 + x = 4$ and $6 + x = 8$ are equivalent equations with solution 2.

inequality A number sentence with a *relation symbol* other than =, such as $>$, $<$, $\geq$, $\leq$, $\neq$, or $\approx$.

integer A number in the set $\{..., -4, -3, -2, -1, 0, 1, 2, 3, 4, ...\}$. A whole number or its *opposite*, where 0 is its own opposite.

Multiplication Property of −1 A property of multiplication that says multiplying any number by −1 gives the opposite of the number. For example, $-1 * 5 = -5$ and $-1 * -3 = -(-3) = 3$. Some calculators apply this property with a $[+/-]$ key that toggles between a positive and negative value in the display.

open sentence A number sentence with one or more variables. An open sentence is neither true nor false. For example, $9 + \_\_ = 15$, $? -24 < 10$, and $7 = x + y$ are open sentences.

opposite of a number *n* A number that is the same distance from zero on the number line as *n*,

but on the opposite side of zero. In symbols, the opposite of a number *n* is *–n*, and, in *Everyday Mathematics*, OPP(*n*). If *n* is a negative number, *–n* is a positive number. For example, the opposite of $-5 = 5$. The sum of a number *n* and its opposite is

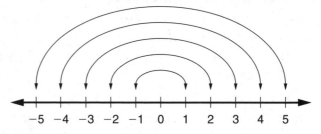

zero; $n + -n = 0$.

order of operations Rules that tell the order in which operations in an expression should be carried out. The conventional order of operations is:

1. Do the operations inside grouping symbols. Work from the innermost set of grouping symbols outward. Inside grouping symbols, follow Rules 2–4.

2. Calculate all the expressions with exponents.

3. Multiply and divide in order from left to right.

4. Add and subtract in order from left to right.

For example:
$$5^2 + (3 * 4 - 2)/5 = 5^2 + (12-2)/5$$
$$= 5^2 + 10/5$$
$$= 25 + 10/5$$
$$= 25 + 2$$
$$= 27$$

reciprocals Two numbers whose product is 1. For example, 5 and $\frac{1}{5}$, $\frac{3}{5}$ and $\frac{5}{3}$, and 0. 2 and 5 are all pairs of multiplicative inverses.

trial-and-error method A method for finding the solution of an equation by trying a sequence of test numbers.

Do-Anytime Activities

To work with your child on concepts taught in this unit, try these interesting and engaging activities:

1. If your child helps with dinner, ask him or her to identify uses of positive and negative numbers in the kitchen. For example, negative numbers might be used to describe the temperature in the freezer. Positive numbers are used to measure liquid and dry ingredients. For a quick game, you might imagine a vertical number line with the countertop as 0; everything above is referenced by a positive number, and everything below is referenced by a negative number. Give your child directions for getting out items by using phrases such as this: "the −2 mixing bowl"; that is, the bowl on the second shelf below the counter.

2. If your child needs extra practice adding and subtracting positive and negative numbers, ask him or her to bring home the directions for the *Credits/Debits Game*. Play a few rounds for review.

3. After your child has completed Lesson 6, ask him or her to explain to you what the following memory device means: *Please Excuse My Dear Aunt Sally.* It represents the rule for the order of operations: parentheses, exponents, multiplication, division, addition, subtraction. Your family might enjoy inventing another memory device that uses the same initial letters; for example, *Please Excuse My Devious Annoying Sibling; Perhaps Everything Might Drop Again Soon,* and so on.

Building Skills Through Games

In Unit 6, your child will work on his or her understanding of algebra concepts by playing games like the ones described below.

Algebra Election See *Student Reference Book*, pages 304 and 305.
Two teams of two players will need 32 *Algebra Election* cards, an Electoral Vote map, 1 six-sided die, 4 pennies or other small counters, and a calculator. This game provides practice with solving equations.

Credits/Debits Game (Advanced Version) See *Student Reference Book*, page 308.
Two players use a complete deck of number cards and a recording sheet to play the advanced version of the *Credits/Debits Game*. This game provides practice with adding and subtracting positive and negative integers.

Top-It See *Student Reference Book*, pages 337 and 338.
Top-It with Positive and Negative Numbers provides practice finding sums and differences of positive and negative numbers. One or more players need 4 each of number cards 0–9 and a calculator to play this *Top-It* game.

As You Help Your Child with Homework

As your child brings assignments home, you might want to go over the instructions together, clarifying them as necessary. The answers listed below will guide you through some of the Unit 6 Study Links.

Study Link 6·1

2. ✓ **3.** ✓ **5.** ✓ **7.** $\frac{1}{19}$

9. $\frac{7}{26}$ **11.** $\frac{3}{4}$ **13.** $12\frac{1}{2}$ lb **14.** $38\frac{1}{4}$ in.

15. $67\frac{1}{2}$ in.^{3} **16.** 81 **17.** -2 **18.** -67

Study Link 6·2

1. $\frac{4}{5}$ **3.** 1 **5.** 1 **7.** $\frac{5}{98}$

9. 10 **10.** 14 **11.** 17 **12.** 13.56

13. 589.36 **14.** 13

Study Link 6·3

1. a. $46 + (-19) = 27$ **c.** $-5 + 6.8 = 1.8$

2. a. -29 **c.** $-2\frac{1}{5}$ **e.** $-3\frac{1}{4}$ **g.** -18.2

3. a. (-2) **c.** $2\frac{1}{4}$ **e.** -3.7 **g.** $-\frac{7}{16}$

4. 2 **5.** 11 **6.** 8 **7.** -6

Study Link 6·4

1. -60 **3.** -6 **5.** -5 **7.** -6

9. $-1,150$ **11.** -54 **13.** -2 **15.** $-\frac{5}{9}$

17. -2 **19. a.** 36 **b.** 77

Study Link 6·6

1. 21 **3.** $\frac{21}{32}$ **5.** 72 **7.** 1

9. 28 **11.** 3 **12.** 23 **13.** 6, 1

14. 2, 1 **15.** 4, 4

Study Link 6·7

1. a. $17 < 27$; $3 * 15 < 100$; $(5 - 4) * 20 = 20$; $12 \neq 12$

b. Sample answer: A number sentence must contain a relation symbol. 56/8 does not include one.

2. a. true **b.** false **c.** false **d.** true

3. a. $(28 - 6) + 9 = 31$ **b.** $20 < (40 - 9) + 11$

c. $(36/6) / 2 < 12$ **d.** $4 * (8 - 4) = 16$

4. a. $60 - 14 = 50$; false **b.** $90 = 3 * 30$; true

c. $21 + 7 < 40$; true **d.** $\sqrt{36} > \frac{1}{2} * 10$; true

5. 0.92 **6.** 3.51 **7.** 251.515

Study Link 6·8

1. a. $b = 19$ **b.** $n = 24$ **c.** $y = 3$ **d.** $m = \frac{1}{5}$

2. a. $\frac{x}{6} = 10$; $x = 60$ **b.** $200 - 7 = n$; $n = 193$

c. $b * 48 = 2{,}928$; $b = 61$

3. Sample answers:

a. $(3 * 11) + (12 - 9)$ **b.** $2 * 18 + 14$

4. 54 **5.** 3.6 **6.** 121

Study Link 6·9

1. 1 **2.** $1\frac{1}{2}$ **3.** 5 **4.** 1

6. Answers vary. **7.** 10 **8.** $\frac{1}{4}$ **9.** $\frac{2}{3}$

10. $\frac{1}{2}$

Study Link 6·10

1. $k - 4 = 5$; $3k - 12 = 15$; $20k - 12 = 15 + 17k$

2. Multiply by 2; M 2
Subtract 3q; S 3q
Add 5; A 5

3. Add 5m; A 5m
Divide by 2; D 2
Subtract 6; S 6

Study Link 6·11

1. $k = 12$ **3.** $x = 1$ **5.** $r = 2$

Study Link 6·12

1. a. $15 \neq 3 * 7$ **b.** $x + 5 = 75$

c. $\frac{9}{9} + 13 \leq 14$

2. a. $200 \div (4 * 5) = 10$

b. $16 + 2^2 - (5 + 3) = 12$

3. a. 46 **b.** 18 **c.** 0 **d.** 8

4. a. $x = -1$ **b.** $y = 6.5$

5. a. Sample answers: -3, $-2\frac{1}{2}$, -2

6. 0.25; 0.21 **7.** 1; 1.28 **8.** 800; 781

Practice with Fractions

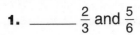

Put a check mark next to each pair of equivalent fractions.

1. _____ $\frac{2}{3}$ and $\frac{5}{6}$

2. _____ $1\frac{3}{4}$ and $\frac{28}{16}$

3. _____ $\frac{24}{30}$ and $\frac{4}{5}$

4. _____ $\frac{7}{3}$ and $\frac{3}{7}$

5. _____ $\frac{56}{8}$ and $\frac{49}{7}$

6. _____ $2\frac{3}{8}$ and $\frac{19}{4}$

Find the reciprocal of each number. Multiply to check your answers.

7. 19 _____

8. $\frac{2}{5}$ _____

9. $3\frac{5}{7}$ _____

10. $\frac{1}{6}$ _____

Multiply. Write your answers in simplest form. Show your work.

11. $\frac{2}{3} * 1\frac{1}{8} =$ _____

12. $3\frac{1}{7} * \frac{7}{22} =$ _____

Solve the number stories.

13. How much does a box containing 5 horseshoes weigh if each horseshoe weighs about $2\frac{1}{2}$ pounds? _____

14. One and one-half dozen golf tees are laid in a straight line, end to end. If each tee is $2\frac{1}{8}$ inches long, how long is the line of tees? _____

15. A standard-size brick is 8 inches long and $2\frac{1}{4}$ inches high and has a depth of $3\frac{3}{4}$ inches. What is the volume of a standard-size brick? _____

Practice

16. $107 + (-82) + 56 =$ _____

17. $4 + (12 + -18) =$ _____

18. $-85 + 66 + (-48) =$ _____

19. $7 + (-11 + -22) =$ _____

STUDY LINK 6·2 | **Fraction Division**

> **Division of Fractions Algorithm**
>
> $$\frac{a}{b} \div \frac{c}{d} = \frac{a}{b} * \frac{d}{c}$$

Divide. Show your work.

1. $\frac{2}{3} \div \frac{5}{6} =$ _____

2. $1\frac{3}{4} \div \frac{28}{16} =$ _____

3. $\frac{24}{30} \div \frac{4}{5} =$ _____

4. $\frac{7}{3} \div \frac{3}{7} =$ _____

5. $\frac{5}{8} \div \frac{5}{8} =$ _____

6. $2 \div \frac{1}{4} =$ _____

7. $\frac{1}{7} \div 2\frac{4}{5} =$ _____

8. $5\frac{5}{6} \div 6 =$ _____

Try This

9. How many $\frac{3}{10}$-centimeter segments are in 3 centimeters? _____ segments

10. How many $\frac{3}{10}$-centimeter segments are in $4\frac{1}{5}$ centimeters? _____ segments

11. How many $\frac{4}{10}$-centimeter segments are in $6\frac{4}{5}$ centimeters? _____ segments

Practice

Round each number to the underlined place.

12. 13.5<u>6</u>1 _____

13. 589.3<u>5</u>52 _____

14. 12.<u>9</u>694 _____

STUDY LINK 6·3 Subtraction of Signed Numbers

For any numbers a and b, $a - b = a + OPP(b)$, or $a - b = a + (-b)$.

1. Rewrite each subtraction problem as an addition problem. Then solve the problem.

a. $46 - 19 =$ _____

b. $-43 - 17 =$ _____

c. $-5 - (-6.8) =$ _____

d. $21 - (-21) =$ _____

2. Subtract.

a. $-72 - (-43) =$ _____

b. _____ $= 4 - (-39)$

c. $-\left(\frac{7}{10}\right) - 1\frac{1}{2} =$ _____

d. $4.8 - (-3.6) =$ _____

e. _____ $= -2\frac{1}{2} - \frac{3}{4}$

f. $-\left(\frac{5}{6}\right) - \left(-\frac{1}{3}\right) =$ _____

g. $-12.3 - 5.9 =$ _____

h. $-8.5 - (-2.7) =$ _____

3. Fill in the missing numbers.

a. $19 = 17 -$ _____

b. $-43 = -26 -$ _____

c. $\frac{1}{2} -$ _____ $= -1\frac{3}{4}$

d. _____ $- \left(-2\frac{4}{5}\right) = 3\frac{7}{10}$

e. $-17.6 =$ _____ $- 13.9$

f. $83.5 = -62.7 -$ _____

g. _____ $= 5\frac{3}{4} - 6\frac{3}{16}$

h. $9.6 -$ _____ $= 10$

Practice

4. $100 = 10^x$; $x =$ _____

5. $10^x = 100$ billion; $x =$ _____

6. 100 million $= 10^x$; $x =$ _____

7. $10^x = 0.00001$; $x =$ _____

STUDY LINK
6·4

*, / of Signed Numbers

A Multiplication Property	A Division Property
◆ The product of two numbers with the same sign is positive.	◆ The quotient of two numbers with the same sign is positive.
◆ The product of two numbers with different signs is negative.	◆ The quotient of two numbers with different signs is negative.

SRB
97

Solve.

1. $-12 * 5 =$ _____

2. $-63 / 7 =$ _____

3. $24 \div (-4) =$ _____

4. $-9 *$ _____ $= 54$

5. $-50 /$ _____ $= 10$

6. $-6 * 5 * 8 =$ _____

7. $48 / (-6 - 2) =$ _____

8. $(-8 * 5) + 12 =$ _____

9. $50 * (-23) =$ _____

10. $6 * (12 + 15) =$ _____

11. $(-90 \div 10) + (-45) =$ _____

12. $56 / (-7) / (-4) =$ _____

13. _____ $* (-7) * (-4) = -56$

14. _____ $\div 40 = -9$

Try This

15. $\frac{2}{3} * \left(-\frac{5}{6}\right) =$ _____

16. $(8 * (-3)) - (8 * (-9)) =$ _____

17. $0.25 * (-8) =$ _____

18. $\left(-\frac{3}{4}\right) \div \left(-\frac{1}{2}\right) =$ _____

19. Evaluate each expression for $b = -7$.

    a. $(-9 * b) - 27 =$ _____

    b. $11 * (-b) =$ _____

    c. $-b / (-14) =$ _____

    d. $b - (b + 16) =$ _____

STUDY LINK 6·5 | **Turn-Around Patterns**

SRB
105

Fill in the missing numbers in the tables. Look for patterns in the results.

1.

x	y	OPP(x)	OPP(y)	$x + y$	$y + x$	$x - y$	$y - x$
7	9	−7	−9	16			
−2	12						
−3	−9						
$\frac{2}{3}$	$\frac{5}{6}$						
2.7	−1.9						
2^2	2^3						

Which patterns did you find in your completed table?

_____

_____

_____

2.

x	y	$\frac{1}{x}$	$\frac{1}{y}$	$x * y$	$y * x$	$x \div y$	$y \div x$
7	9	$\frac{1}{7}$	$\frac{1}{9}$	63			
−2	12						
−3	−9						
$\frac{2}{3}$	$\frac{5}{6}$						
2.7	−1.9						
2^2	2^3						

Which patterns did you find in your completed table?

_____

_____

_____

STUDY LINK 6·6 Using Order of Operations

Please Excuse My Dear Aunt Sally
Parentheses Exponents Multiplication Division Addition Subtraction

Evaluate each expression.

1. $5 + 6 * 3 - 2 =$ _____

2. $4 * 9 / 2 + (-4 + 6) =$ _____

3. $\frac{1}{2} + \frac{5}{8} * \frac{1}{2} \div 2 =$ _____

4. $(2.3 + 7.8) * 4 + 3 =$ _____

5. $4^2 + 7(3 - (-5)) =$ _____

6. $((2 * 4) + 3) * 6 / 2 =$ _____

Evaluate the following expressions for $m = -3$.

7. $-\frac{m}{m} + 6 - 4 =$ _____

8. $((4 + 11) * -3) / 9 * (-m) =$

9. $m^2 + (-(m^3)) - 8 =$ _____

10. $\frac{1}{2} * m \div \frac{5}{4} + \frac{3}{5} - \frac{1}{10} =$ _____

Practice

Find each missing number.

11. 3 gal 7 qt = 4 gal _____ qt

12. 5 gal 3 qt = _____ qt

13. 13 pt = _____ qt _____ pt

14. 10 c = _____ qt _____ pt

15. 18 qt = _____ gal _____ pt

Units of Capacity

2 cups (c) = 1 pint (pt)

2 pints = 1 quart (qt)

4 quarts = 1 gallon (gal)

145

| **Number Sentences**

1. a. Draw a circle around each number sentence.

$17 < 27$ $3 * 15 < 100$ $56 / 8$

$(5 - 4) * 20 = 20$ $(4 + 23) / 9$ $12 \neq 12$

b. Choose one item that you did not circle. Explain why it is not a number sentence.

_____

_____

2. Tell whether each number sentence is true or false.

a. $9 - (6 + 2) > 0.5$ _____

b. $94 = 49 - 2 * 2$ _____

c. $\frac{24}{6} < 33 / 11$ _____

d. $70 - 25 = 45$ _____

3. Insert parentheses to make each number sentence true.

a. $28 - 6 + 9 = 31$

b. $20 < 40 - 9 + 11$

c. $36 / 6 / 2 < 12$

d. $4 * 8 - 4 = 16$

4. Write a number sentence for each word sentence. Tell whether the number sentence is true or false.

Word sentence	Number sentence	True or false?
a. If 14 is subtracted from 60, the result is 50.	_____	_____
b. 90 is 3 times as much as 30.	_____	_____
c. 21 increased by 7 is less than 40.	_____	_____
d. The square root of 36 is greater than half of 10.	_____	_____

Practice

5. $1.867 - 0.947 =$ _____ **6.** $6 - 2.49 =$ _____ **7.** $256.3 - 4.785 =$ _____

STUDY LINK 6·8 Solving Simple Equations

1. Find the solution to each equation.

   a. $b - 7 = 12$ _____ **b.** $53 = n + 29$ _____

   c. $45 / y^2 = 5$ _____ **d.** $m * \frac{2}{3} = 1 - \frac{13}{15}$ _____

2. Translate the word sentences below into equations. Then solve each equation.

Word sentence	Equation	Solution
a. If you divide a number by 6, the result is 10.	_____	_____
b. Which number is 7 less than 200?	_____	_____
c. A number multiplied by 48 is equal to 2,928.	_____	_____
d. 27 is equal to 13 increased by which number?	_____	_____

3. For each problem, use parentheses and as many numbers and operations as you can to write an expression equal to the target number. You may use each number only once in an expression. Write expressions with more than two numbers.

   a. Numbers: 3, 9, 11, 12, 19 Target number: 36 _____

   b. Numbers: 1, 2, 6, 14, 18 Target number: 50 _____

   c. Numbers: 4, 5, 8, 14, 17 Target number: 22 _____

   d. Numbers: 6, 7, 12, 14, 20 Target number: 41 _____

Practice

Complete.

4. $540 \div 90 =$ _____ $\div 9$ 5. $36 \div 6 =$ _____ $\div 0.6$ 6. _____ $\div 11 = 1.21 \div 0.11$

STUDY LINK 6·9 | # Solving Pan-Balance Problems

Solve these pan-balance problems. In each figure, the two pans are balanced.

1. One ball weighs

as much as _____ coin(s).

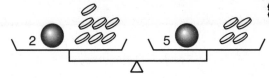

2. One cube weighs

as much as _____ marble(s).

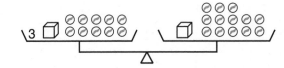

3. One *x* weighs

as much as _____ *y*(s).

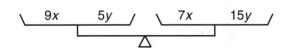

4. One *a* weighs

as much as _____ *b*(s).

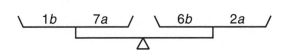

Make up two pan-balance problems for a classmate to solve.

5. _____

_____

6. _____

_____

Practice

7. $605 * \frac{1}{10} = 605 \div$ _____

8. $72 *$ _____ $= 72 \div 4$

9. _____ $* 30 = (2 * 30) \div 3$

10. _____ $* (x + 5) = \frac{x + 5}{2}$

151

STUDY LINK 6·10 | Balancing Equations

For Problem 1, record the result of each operation on each pan.

1. Original pan-balance equation

Operation	
(in words)	**(abbreviation)**
Subtract 4.	S 4
Multiply by 3.	M 3
Add 17k.	A 17k

$k = 9$

$ = $

$ = $

$ = $

For Problems 2 and 3, record the operation that was used to obtain the result on each pan balance.

2. Original pan-balance equation

Operation	
(in words)	**(abbreviation)**
_____	_____
_____	_____
_____	_____

$1 + 1.5q = 2q - 2.5$

$2 + 3q = 4q - 5$

$2 = q - 5$

$7 = q$

3. Original pan-balance equation

Operation	
(in words)	**(abbreviation)**
_____	_____
_____	_____
_____	_____

$-3m + 12 = 13 - 5m$

$2m + 12 = 13$

$m + 6 = 6\frac{1}{2}$

$m = \frac{1}{2}$

153

STUDY LINK 6·11 | **Solving Equations**

Solve each equation. Then check the solution.

1. $9 + 5k = 45 + 2k$

Original equation

$$\underline{\quad 9 + 5k = 45 + 2k \quad}$$

Operation

$$\underline{S\ 9} \qquad \underline{5k = 36 + 2k}$$

_____

$$\underline{\qquad k = 12 \qquad}$$

Check

_____

2. $\frac{9}{2}m - 8 = -5.5 + 4m$

Original equation

_____

Operation

_____   _____

_____   _____

_____   _____

Check

_____

3. $24x - 10 = 18x - 4$

Original equation

_____

Operation

_____   _____

_____   _____

_____   _____

Check

_____

4. $12d - 9 = 15d + 9$

Original equation

_____

Operation

_____   _____

_____   _____

_____   _____

Check

_____

5. $-6r - 5 = 7 - 12r$

Original equation

_____

Operation

_____   _____

_____   _____

_____   _____

Check

_____

6. $\frac{1}{3}p + 7 = 12 - \frac{2}{3}p$

Original equation

_____

Operation

_____   _____

_____   _____

_____   _____

Check

_____

155

STUDY LINK 6·12 | **Review**

1. Write a number sentence for each word sentence.

Word sentence	Number sentence

   a. 15 is not equal to 3 times 7. _____

   b. 5 more than a number is 75. _____

   c. 13 more than 9 divided by 9 is
   less than or equal to 14. _____

2. Insert parentheses to make each equation true.

   a. $200 \div 4 * 5 = 10$ **b.** $16 + 2^2 - 5 + 3 = 12$

3. Use the order of operations to evaluate each expression.

   a. $5 * 6 + 8 * 2 =$ _____   **b.** $20 - \dfrac{8}{2^2} =$ _____

   c. $40 + 8 - 24 * 2 =$ _____   **d.** $4^2 \div (4 * 2) + 3 * 2 =$ _____

4. Solve each equation.

   a. $3x - 5 = 5x - 3$ **b.** $\dfrac{(4y + 5)}{2} = y + 9$

   Solution _____   Solution _____

5. Name three solutions of the inequality. Then graph the solution set.

   a. $f < -\dfrac{3}{2}$

   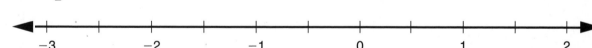

Practice

6. $2.52 ÷ 12 Estimate _____ Quotient _____

7. 45)57.60   Estimate _____ Quotient _____

8. 120)93,720 Estimate _____ Quotient _____

STUDY LINK
6·13

Unit 7: Family Letter

Probability and Discrete Mathematics

All of us are aware that the world is filled with uncertainties. As Ben Franklin wrote, "Nothing is certain except death and taxes!" Of course, there are some things we can be sure of: The sun will rise tomorrow, for example. We also know that there are degrees of uncertainty—some things are more likely to happen than others. There are occurrences that, although uncertain, can be predicted with reasonable accuracy.

While predictions are usually most reliable when they deal with general trends, it is possible and often helpful to predict the outcomes of specific situations. In Unit 7, your child will learn how to simulate a situation with random outcomes and how to determine the likelihood of various outcomes. Additionally, the class will analyze games of chance to determine whether or not they are fair; that is, whether or not all players have the same chance of winning.

We will be looking at two tools for analyzing probability situations: tree diagrams (familiar from single-elimination sports tournaments) and Venn diagrams (circle diagrams that show relationships between overlapping groups).

One lesson concerns strategies for taking multiple-choice tests based on probability. Should test-takers guess at answers they don't know? Your child will learn some of the advantages and disadvantages of guessing on this type of test.

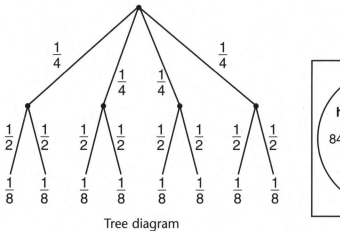

Tree diagram

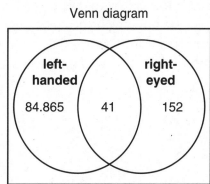

Venn diagram

Please keep this Family Letter for reference as your child works through Unit 7.

159

Vocabulary

Important terms in Unit 7:

equally likely outcomes *Outcomes* of a chance experiment or situation that have the same probability of happening. If all the possible outcomes are equally likely, then the probability of an event is equal to:

$$\frac{\text{number of favorable outcomes}}{\text{number of possible outcomes}}$$

expected outcome The average outcome over a large number of repetitions of a random experiment. For example, the expected outcome of rolling one die is the average number of dots showing over a large number of rolls.

outcome A possible result of a chance experiment or situation. For example, heads and tails are the two possible outcomes of tossing a coin.

probability A number from 0 through 1, giving the likelihood that an event will happen. The closer a probability is to 1, the more likely the event is to happen.

probability tree diagram A drawing used to analyze a *probability* situation that consists of two or more choices or stages. For example, the branches of the probability tree diagram below represent the four *equally likely outcomes* when one coin is flipped two times.

random number A number produced by a random experiment, such as rolling a die or spinning a spinner. For example, rolling a fair die produces random numbers because each of the six possible numbers 1, 2, 3, 4, 5, and 6 has the same chance of coming up.

simulation A model of a real situation. For example, a fair coin can be used to simulate a series of games between two equally matched teams.

Venn diagram A picture that uses circles or rings to show relationships among sets. The Venn diagram below shows the number of students who have a dog, a cat, or both.

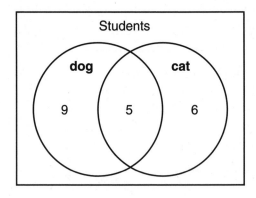

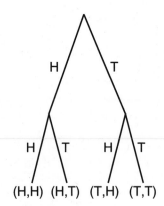

Do-Anytime Activities

To work with your child on the concepts taught in this unit and in previous units, try these interesting and rewarding activities:

1. While playing a game that uses a die, keep a tally sheet of how many times a certain number lands. For example, try to find out how many times during the game the number 5 comes up. Have your child write the probability for the chosen number. ($\frac{1}{6}$ is the probability that any given number on a six-sided die will land.) The tally sheet should show how many times the die was rolled during the game and how many times the chosen number came up.

2. Have your child listen to the weather forecast on television and pick out the language of probability. Have him or her listen for such terms as *likely, probability, (percent) chance, unlikely,* and so on.

3. Watch with your child for events that occur without dependence on any other event. In human relationships, truly independent events may be difficult to isolate, but this observation alone helps to define the randomness of events. Guide your child to see the difference between dependent events and independent events. For example, "Will Uncle Mike come for dinner?" depends on whether or not he got his car fixed. However, "Will I get heads or tails when I flip this coin?" depends on no other event.

Building Skills through Games

In Unit 7, your child will continue to review concepts from previous units and prepare for topics in upcoming units by playing games such as:

***2–4–8 and 3–6–9 Frac–Tac–Toe* (Percent Versions)** See *Student Reference Book,* pages 314–316
The two versions, *2-4-8 Frac-Tac-Toe* and *3-6-9 Frac-Tac-Toe,* help students practice conversions between fractions and percents. Two players need a deck of number cards with four each of the numbers 0–10; a gameboard, a 5 × 5 grid that resembles a bingo card; a *Frac-Tac-Toe* Number-Card Board; markers or counters in two different colors, and a calculator.

Angle Tangle See *Student Reference Book,* page 306
Two players need a protractor, straightedge, and blank sheets of paper to play this game. Mastering the estimation and measurement of angles is the goal of *Angle Tangle.*

Name That Number See *Student Reference Book,* page 329
This game provides practice in using order of operations to write number sentences. Two or three players need a complete deck of number cards.

Solution Search See *Student Reference Book,* page 332
This game provides practice solving open number sentences. Players use a complete deck of number cards as well as *Solution Search* cards to solve inequalities.

As You Help Your Child with Homework

As your child brings assignments home, you may want to go over the instructions together, clarifying them as necessary. The answers listed below will guide you through some of the Unit 7 Study Links.

Study Link 7·1

1. Quarter, nickel, dime; No. There is an unequal number of each type of coin.

2. 1, 2, 4, 5, 10, and 20

   Yes. Each number card is a factor of 20.

3. 37.5%   4. 100%   5. 25%, 50%, 75%

6. 27.12

Study Link 7·2

1. No. Sample answer: Teams should be evenly matched. A team selected at random might not have a balance of skilled and unskilled players.

2. Yes and no. Sample answer: In an elementary school, preference for the better seats should go to the youngest children so they can see the game. However, in Grades 3–6, the principal should choose seat assignments randomly.

3. Disagree. Sample answer: There is always an even chance of this spinner landing on black or white. Previous spins do not affect the outcome.

4. Agree. Sample answer: There is always a better chance that this spinner will land on white because the white area is larger. The outcome does not depend on previous spins.

Study Link 7·3

1. 6 ways   2. 30, 26, 23, 22, 19, 18, 16, 15, 12, 9

3a. 25%   3b. 33.33%

Study Link 7·4

3. 12   4. 15   5. 15

Study Link 7·5

1. Tree diagram probabilities (from top, left to right)
   $\frac{1}{2}, \frac{1}{2}$

   Box 1: $\frac{1}{3}, \frac{1}{3}, \frac{1}{3}, \frac{1}{3}, \frac{1}{3}, \frac{1}{3}$

   Box 2: $\frac{1}{2}, \frac{1}{2}, \frac{1}{2}, \frac{1}{2}, \frac{1}{2}, \frac{1}{2}, \frac{1}{2}, \frac{1}{2}, \frac{1}{2}, \frac{1}{2}, \frac{1}{2}, \frac{1}{2}$

   Box 3: $\frac{1}{12}, \frac{1}{12}, \frac{1}{12}, \frac{1}{12}, \frac{1}{12}, \frac{1}{12}, \frac{1}{12}, \frac{1}{12}, \frac{1}{12}, \frac{1}{12}, \frac{1}{12}, \frac{1}{12}$

2. 12

3. a. $\frac{1}{6}$ b. $\frac{3}{3}$, or 100%   c. $\frac{1}{3}$ d. 0%

4. 36.5   5. 22.6   6. 12.6

Study Link 7·6

1. a. Track   b. Basketball   c. 22   d. 8

   e. 30   f. 52   g. 22

3. $\frac{17}{40}$ 4. $2\frac{11}{12}$ 5. $8\frac{3}{20}$

Study Link 7·7

1. Tree diagram probabilities (from top, left to right)
   $\frac{1}{4}, \frac{1}{4}, \frac{1}{4}, \frac{1}{4}$

   R1: $\frac{1}{3}, \frac{1}{3}, \frac{1}{3}$; **R2:** $\frac{1}{3}, \frac{1}{3}, \frac{1}{3}$; **R3:** $\frac{1}{3}, \frac{1}{3}, \frac{1}{3}$; **G:** $\frac{1}{3}, \frac{1}{3}, \frac{1}{3}$

   Bottom row probabilities:

   $\frac{1}{12}, \frac{1}{12}, \frac{1}{12}, \frac{1}{12}, \frac{1}{12}, \frac{1}{12}, \frac{1}{12}, \frac{1}{12}, \frac{1}{12}, \frac{1}{12}, \frac{1}{12}, \frac{1}{12}$

   a. 50   b. 25

2. a. HHT; HTH; HTT; THH; THT; TTH

   b. 37.5   c. 87.5

Study Link 7·8

1. C, D, A, B

2. Tree diagram with branches labeled as follows (from left to right):

   Swimsuits: red, white, blue

   Sandals: red, white; red, white; red, white

   a. 6   b. $\frac{2}{6}$, or $\frac{1}{3}$

3. a.   b. 25

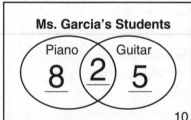

Ms. Garcia's Students

Piano 8 (2) 5 Guitar

10

Sample answer: 8 students play the piano, 5 students play the guitar, 2 students play both instruments, and 10 students play neither instrument. 8 + 2 + 5 + 10 = 25

STUDY LINK 7·1 | Outcomes and Probabilities

SRB
150–153

Complete the table.

Experiment	Possible Outcomes	Outcomes Equally Likely?
Example: Spin the spinner. A | B C	A, B, C	No. The area for C is twice as large as each of the other 2 areas.
1. Choose a coin. D Q Q N D D D Q		
2. Choose a factor of 20. 1 20 10 2 4 5		

Use the problems from the table to answer the following questions.
Express each probability as a percent.

3. What is the probability of selecting a quarter from the coins in Problem 1? _____

4. What is the probability of choosing a factor of 20 from the cards in Problem 2? _____

5. Suppose you spin the spinner from the Example in the table. Complete the number sentence below to determine the probability of the spinner landing on A or C.

_____ + _____ = _____
Probability of A   Probability of C   Probability of A or C

Practice

Simplify the expression using the order of operations.

6. $3.8 + 6.4 \div 0.2 - 1.8 * 2.6 - 3.2 \div 0.8$ _____

STUDY LINK 7·2 | **Using Random Numbers**

1. A gym teacher is dividing her class into two teams to play soccer.
   Do you think she should choose the teams at random? _____

   Explain. _____

   _____

   _____

2. The entire school is going to a baseball game. Some seats are better
   than others. Should the principal select the section where each class
   will sit at random? _____

   Explain. _____

   _____

   _____

   _____

3. The spinner at the right has landed on black 5 times in a row. Renee
   says, "On the next spin, the spinner is more likely to land on white
   than on black."

   Do you agree or disagree with Renee? _____

   Explain. _____

   _____

   _____

4. The spinner at the right has landed on black 5 times in row. Matthew
   says, "On the next spin, the spinner has a better chance of landing on
   white than on black."

   Do you agree or disagree with Matthew? _____

   Explain. _____

   _____

   _____

   _____

STUDY LINK 7·3   Making Organized Lists

Solve each problem by making an organized list. The list in Problem 1 has been started for you.

SRB
156

1. In how many ways can you make $0.60 using at least 1 quarter? You can only use quarters, dimes, and nickels.

_____

Q	D	N
1	3	1

2. You throw three darts and hit the target at the right. List the different total points that are possible.

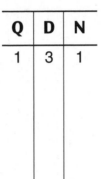

10 pts	6 pts	3 pts	Total pts

Use what you know about angle measures of sectors to find the probabilities in Problem 3.

Example:

Probability of landing on striped sector $= \frac{150°}{360°} = \frac{5}{12} = 41.67\%$

3. Find the probability of the spinner landing on

a. white. _____

b. black. _____

167

STUDY LINK 7·4

Lists and Tree Diagrams

Suppose members of the hiking club are served a breakfast bag whenever they have a Saturday morning meeting. Members use the form at the right to place their orders.

Breakfast Order Form

Beverage
☐ Milk   ☐ Water

Bagel
☐ Plain ☐ Raisin

Fruit
☐ Apple ☐ Banana ☐ Orange

1. Complete the organized list of the possible breakfast bags.

Beverage	Bagel	Fruit
M	P	A
M	P	B

Beverage	Bagel	Fruit
W	P	A
W	P	B

2. Use your organized list to complete the tree diagram.

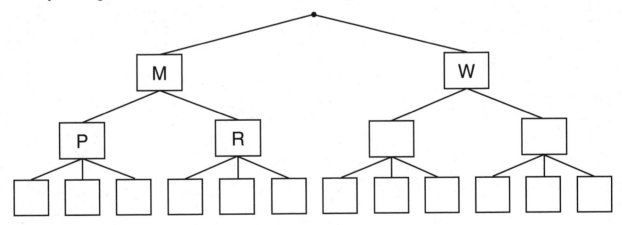

3. How many different breakfast bags are possible? _____

4. Suppose 60 members fill out an order form. About how many people would you expect to order milk and a plain bagel?   _____ people

5. Suppose each of the 60 members brings 2 guests to the next Saturday meeting. About how many people would you expect to order water, a raisin bagel, and an orange?   _____ people

169

STUDY LINK 7·5 | **A Random Draw and a Tree Diagram**

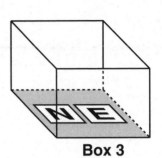

SRB
154 155

Boxes 1, 2, and 3 contain letter tiles.

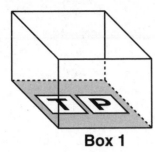

Box 1   Box 2   Box 3

Suppose you draw one letter from each box without looking. You lay the letters in a row—the Box 1 letter first, the Box 2 letter second, and the Box 3 letter third.

1. Complete the tree diagram. Fill in the blanks to show the probability for each branch.

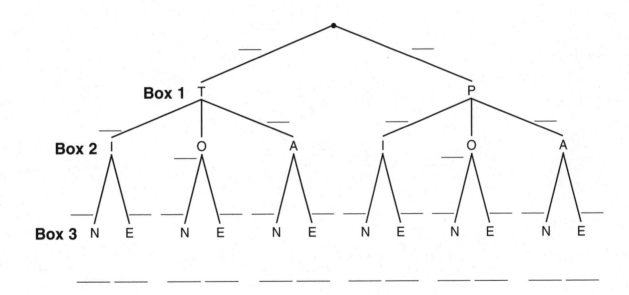

2. How many possible combinations of letter tiles are there? _____

3. What is the probability of selecting:

   a. the letters P and I? _____   **b.** the letter I, O, or A? _____

   c. the letter combinations TO or PO? _____   **d.** two consonants in a row? _____

| **Practice** |

4. 657 ÷ 18 = _____   **5.** 858.8 ÷ 38 = _____   **6.** 1,575 ÷ 125 = _____

171

STUDY LINK
7·6
Venn Diagrams

There are 200 girls at Washington Middle School.

◆ 30 girls are on the track team.

◆ 38 girls are on the basketball team.

◆ 8 girls are on both teams.

1. Complete the Venn diagram below to show the number of girls on each team.

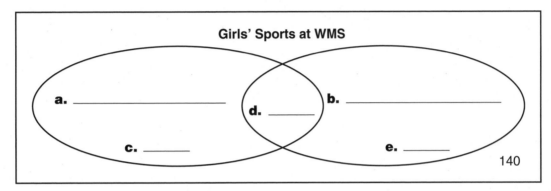

Girls' Sports at WMS

a. _____

b. _____

d. _____

c. _____

e. _____

140

   f. How many girls are on one team but not both? _____ girls

   g. How many girls are on the track team but not the basketball team? _____ girls

2. Write a situation (2d) for the Venn diagram below. Complete the diagram by adding a title (2a) and labeling each ring (2b and 2c).

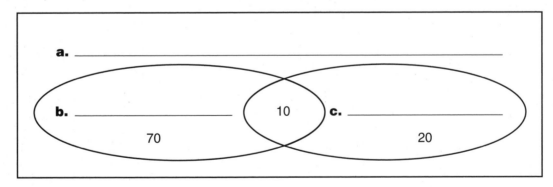

a. _____

b. _____

10

c. _____

70

20

d. _____

_____

_____

Practice

3. $\frac{7}{8} - \frac{9}{20} =$ _____

4. $7\frac{1}{3} - 4\frac{5}{12} =$ _____

5. $9\frac{2}{5} - 1\frac{1}{4} =$ _____

173

More Tree Diagrams

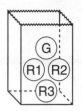

Denise has 3 red marbles and 1 green marble in a bag. She draws
1 marble at random. Then she draws a second marble without putting
the first marble back in the bag.

1. Find the probabilities for each branch of the tree diagram below.

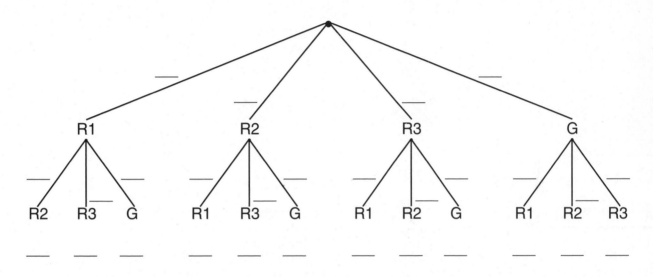

    a. What is the probability that Denise will select 2 red marbles? _____%

    b. What is the probability that Denise will first
        draw a green marble and then a red marble? _____%

2. Three coins are tossed.

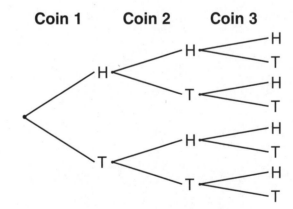

Outcomes
HHH
TTT

    a. Complete the table of possible outcomes at the right.

    b. What is the probability of tossing *exactly* 2 HEADS? _____%

    c. What is the probability of tossing *at least* 1 TAIL? _____%

STUDY LINK 7·8 Reviewing Probability

1. Each fraction in the left column below shows the probability of a chance event. Write the letter of the description next to the fraction that represents it.

_____ $\frac{1}{3}$ **A.** Probability of getting HEADS if you flip a coin

_____ $\frac{1}{4}$ **B.** Probability of rolling 3 on a 6-sided die

_____ $\frac{1}{2}$ **C.** Probability of choosing a red ball from a bag containing 2 red balls, 3 white balls, and 1 green ball

_____ $\frac{1}{6}$ **D.** Probability of drawing a heart card from a deck of playing cards

2. Sidone bought 3 new swimsuits— 1 red suit, 1 blue suit, and 1 white suit. She also bought 2 pairs of beach sandals—1 red pair and 1 white pair. Make a tree diagram in the space at the right to show all possible combinations of swimsuits and sandals.

   a. How many different combinations of suits and sandals are there? _____

   b. If Sidone chooses a swimsuit and a pair of sandals at random, what is the probability that they will be the same color? _____

3. **a.** Ten students in Ms. Garcia's class play the piano. Seven students play the guitar. Two students play both the piano and the guitar. Complete the Venn diagram at the right.

   b. How many students are in Ms. Garcia's class? _____ students

   Explain how you know.

   _____

   _____

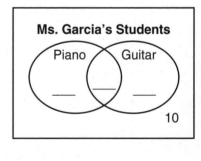

Ms. Garcia's Students

Piano   Guitar

10

177

STUDY LINK 7·9

Unit 8: Family Letter

Rates and Ratios

The next unit is devoted to the study of rates and ratios. Fraction and decimal notation will be used to express rates and ratios and to solve problems.

Ratios compare quantities that have the same unit. These units cancel each other in the comparison, so the resulting ratio has no units. For example, the fraction $\frac{2}{20}$ could mean that 2 out of 20 people in a class got an A on a test or that 20,000 out of 200,000 people voted for a certain candidate in an election.

Another frequent use of ratios is to indicate relative size. For example, a picture in a dictionary drawn to $\frac{1}{10}$ scale means that every length in the picture is $\frac{1}{10}$ the corresponding length in the actual object. Students will use ratios to characterize relative size as they examine map scales and compare geometric figures.

Rates, on the other hand, compare quantities that have different units. For example, rate of travel, or speed, may be expressed in miles per hour (55 mph); food costs may be expressed in cents per ounce (17 cents per ounce) or dollars per pound ($2.49 per pound).

Easy ratio and rate problems can be solved intuitively by making tables, such as *What's My Rule?* tables. Problems requiring more complicated calculations are best solved by writing and solving proportions. Students will learn to solve proportions by cross multiplication. This method is based on the idea that two fractions are equivalent if the product of the denominator of the first fraction and the numerator of the second fraction is equal to the product of the numerator of the first fraction and the denominator of the second fraction. For example, the fractions $\frac{4}{6}$ and $\frac{6}{9}$ are equivalent because $6 * 6 = 4 * 9$. This method is especially useful because proportions can be used to solve any ratio and rate problem. It will be used extensively in algebra and trigonometry.

$$9 * 4 = 36 \qquad\qquad 6 * 6 = 36$$

$$\frac{4}{6} = \frac{6}{9}$$

Students will apply these rate and ratio skills as they explore nutrition guidelines. The class will collect nutrition labels and design balanced meals based on recommended daily allowances of fat, protein, and carbohydrate. You might want to participate by planning a balanced dinner together and by examining food labels while shopping with your child. Your child will also collect and tabulate various kinds of information about your family and your home and then compare the data by converting them to ratios. In a final application lesson, your child will learn about the Golden Ratio—a ratio found in many works of art and architecture.

179

Vocabulary

Important terms in Unit 8:

Golden Ratio The *ratio* of the length of the long side to the length of the short side of a Golden Rectangle, approximately equal to 1.618 to 1. The Greek letter ϕ (phi) sometimes stands for the Golden Ratio. The Golden Ratio is an irrational number equal to $\frac{1 + \sqrt{5}}{2}$.

n-to-1 ratio A *ratio* of a number to 1. Every ratio *a:b* can be converted to an *n*-to-1 ratio by dividing *a* by *b*. For example, a ratio of 3 to 2 is a ratio of 3 / 2 = 1.5, or a 1.5-to-1 ratio.

part-to-part ratio A *ratio* that compares a part of a whole to another part of the same whole. For example, *There are 8 boys for every 12 girls* is a part-to-part ratio with a whole of 20 students. Compare to *part-to-whole ratio*.

part-to-whole ratio A *ratio* that compares a part of a whole to the whole. For example, *8 out of 20 students are boys* and *12 out of 20 students are girls* are part-to-whole ratios. Compare to *part-to-part ratio*.

per-unit rate A *rate* with 1 unit of something in the denominator. Per-unit rates tell how many of one thing there are for one unit of another thing. For example, *3 dollars per gallon, 12 miles per hour,* and *1.6 children per family* are per-unit rates.

proportion A number sentence equating two fractions. Often the fractions in a proportion represent *rates* or *ratios*.

rate A comparison by division of two quantities with different units. For example, traveling 100 miles in 2 hours is an average rate of $\frac{100 \text{ mi}}{2 \text{ hr}}$ or 50 miles per hour. Compare to *ratio*.

ratio A comparison by division of two quantities with the same units. Ratios can be fractions, decimals, percents, or stated in words. Ratios can also be written with a colon between the two numbers being compared. For example, if a team wins 3 games out of 5 games played, the ratio of wins to total games is $\frac{3}{5}$, 3 / 5, 0.6, 60%, 3 to 5, or 3:5 (read "three to five"). Compare to *rate*.

similar figures Figures that have the same shape, but not necessarily the same size. For example, all squares are similar to one another, and the preimage and image of a *size-change* are similar. The *ratio* of lengths of corresponding parts of similar figures is a *scale* or *size-change factor*. In the example below, the lengths of the sides of the larger polygon are 2 times the lengths of the corresponding sides of the smaller polygon. Compare to congruent.

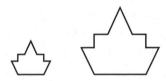

Similar polygons

size-change factor Same as *scale factor*.

scale factor (1) The *ratio* of lengths on an image and corresponding lengths on a preimage in a *size-change*. Same as *size-change factor*. (2) The *ratio* of lengths in a scale drawing or scale model to the corresponding lengths in the object being drawn or modeled.

Do-Anytime Activities

To work with your child on the concepts taught in this unit and in previous units, try these interesting and rewarding activities:

1. Look with your child through newspapers and magazines for photos, and check them to see if a size-change factor is mentioned in the caption: that is, 2X for an enlarged photo 2 times life-size; or $\frac{1}{2}$X for a photo reduced by half. You might find photos of insects, stars, bacteria, and so on. Have your child explain to you what the size-change factor means.

2. Encourage your child to read nutrition labels and calculate the percent of fat in the item.

$$\frac{\text{fat calories}}{\text{total calories}} = \frac{?}{100} = ?\% \text{ of calories from fat}$$

   If your child enjoys this activity, extend it by figuring the percent of calories from protein and carbohydrate.

3. Help your child distinguish between part-to-part and part-to-whole ratios. When comparing a favorite sports team's record, decide which ratio is being used. For example, wins to losses (such as 5 to 15) or losses to wins (15 to 5) are part-to-part ratios. Part-to-whole ratios are used to compare wins to all games played (5 out of 20) or losses to all games played (15 out of 20).

Building Skills through Games

In Unit 8, your child will continue to review concepts from previous units and prepare for topics in upcoming units by playing games such as:

***Division Top-It* (Advanced Version)** See *Student Reference Book,* page 336
Two to four people can play this game using number cards 1–9. Players apply place-value concepts, division facts, and estimation strategies to generate whole-number division problems that will yield the largest quotient.

Spoon Scramble See *Student Reference Book,* page 333
Playing *Spoon Scramble* helps students practice finding fraction, decimal, and percent parts of a whole. Four players need a deck of 16 *Spoon Scramble* cards and 3 spoons to play this game.

As You Help Your Child with Homework

As your child brings assignments home, you may want to go over the instructions together, clarifying them as necessary. The answers listed below will guide you through some of the Study Links in this unit.

Study Link 8·1

2. a. 13   **b.** $6.50

3. a.

Words	1	2	4	5
Minutes	75	150	300	375

  b. 375 words   **c.** 14 minutes

4. 0.6 hours   **5.** About 44

6. −1   **7.** −1.856

Study Link 8·2

1. $\frac{3}{15} = \frac{a}{125}$; 25

2. $\frac{30}{48} = \frac{w}{64}$; 40

3. $\frac{240}{10} = \frac{216}{g}$; 9

4. $\frac{60.96}{2} = \frac{c}{3}$; 91.44

  Sample estimates are given.

5. 600; 674   **6.** 100; 91   **7.** 40; 35

Study Link 8·3

1. a. $0.13 per worm   **b.** $3.38

2. a. $0.18 per oz   **b.** $2.88

3. 150,000 people   **4.** 625 gallons

5. $840; $15,120   **6.** $\frac{1}{2}$ cent

7. 16 hours; Sample answer: 128 oz = 1 gal; 12 gal = 1,536 oz; $\frac{1,536 \text{ oz}}{1.6 \text{ oz per min}}$ = 960 min; $\frac{960 \text{ min}}{60 \text{ min per hour}}$ = 16 hr

Study Link 8·4

Answers vary.

Study Link 8·5

Answers vary.

Study Link 8·6

1. 25   **2.** 27   **3.** 24; 40

4. San Miguel Middle School; Sample answer: I wrote a ratio comparing the number of students to the number of teachers for each school. Richards Middle School, $\frac{14}{1}$; San Miguel, $\frac{13}{1}$.

5.

Shelf	Mystery Books	Adventure Books	Humor Books
1	4	10	18
2	6	15	27
3	8	20	36
4	10	25	45
5	12	30	54
6	14	35	63

6. 14.83   **7.** 88.43   **8.** 12.06

Study Link 8·7

1. 20   **2.** 57   **3.** 27   **4.** 6

5. 250   **6.** 42   **7.** $12\frac{1}{24}$ **8.** $2\frac{8}{9}$

9. $4\frac{11}{20}$ **10.** $3\frac{27}{40}$

Study Link 8·8

Answers vary for 5a and 5b.

5. a. $6\frac{1}{2}$ in.; $4\frac{3}{4}$ in.   **b.** 5 in.; 3 in.   **c.** $7\frac{1}{4}$ in.; $3\frac{3}{4}$ in.   **d.** $9\frac{1}{2}$ in.; $4\frac{1}{4}$ in.   **e.** 11 in.; $8\frac{1}{2}$ in.

6. Answers vary.

7. Sample answers: **a.** $6\frac{1}{2}$ **b.** 11

8. 2.3   **9.** 57.7   **10.** 10.2

Study Link 8·9

1. a. 64 mm   **b.** 32 mm

2. a. 45 mm   **b.** 180 mm; $\frac{1}{4}$

3. a. 45 mm   **b.** 15 mm; 3

4. a. 55 mm   **b.** 165 mm; $\frac{1}{3}$

Study Link 8·10

1. a. 2:1   **b.** 90°   **c.** 9   **d.** 2:1

2. a. 15   **b.** $\frac{3}{2}$ **3.** 90

4. 0.007   **5.** 63.498   **6.** 4.892   **7.** 5.920

Study Link 8·11

1. 1.2; Answers vary.

2. 1.65; No. Sample answer: The ratio for a standard sheet of paper is about 1.3 to 1.

3. Lucille; Sample answer: Compare ratios of correct problems to total problems. Jeffrey's ratio is 0.93 to 1; Lucille's ratio is 0.94 to 1.

4. 12   **5.** 2.8; Answers vary.   **6.** 888

7. 21,228   **8.** 15,456   **9.** 126,542

Study Link 8·12

1. a. 3.14 to 1   **b.** 1.6 to 1   **c.** 2 to 1   **d.** 1 to 1   **e.** 3 to 5

2. a. 40%   **b.** 3:5, or $\frac{3}{5}$

3. b. $7.50   **c.** 8 cans

4. a. 24 members   **b.** $\frac{3}{5} = \frac{12}{n}$; 20 free throws

5. Answers vary.   **6.** Answers vary.

STUDY LINK 8·1 More Rates and Proportions

1. Bring nutrition labels from a variety of food packages and cans to class. A sample label is shown at the right.

Nutrition Facts
Serving Size 1 slice (23 g)
Servings Per Container 20

Amount Per Serving
Calories 65   Calories from Fat 9

% Daily Value

Total Fat 1 g   2%

Total Carbohydrate 12 g   4%

Protein 2 g

2. Express each rate as a per-unit rate.

a. 143 players per 11 teams [____] players/team

b. $260 for 40 hours [____]/hour

3. Kendis types 150 words in 2 minutes.

minutes	1	2	4	5
words	____	150	____	____

a. Fill in the rate table.

b. At this rate, how many words can Kendis type in 5 minutes? _____

Complete the proportion to show your solution. $\dfrac{[\quad] \text{ words}}{[\quad] \text{ minute}} = \dfrac{[\quad] \text{ words}}{[\quad] \text{ minutes}}$

c. How many minutes would it take Kendis to type 1,050 words? _____

Complete the proportion to show your solution. $\dfrac{[\quad] \text{ words}}{[\quad] \text{ minute}} = \dfrac{[\quad] \text{ words}}{[\quad] \text{ minutes}}$

Try This

Use any method you wish to solve the following problems.

4. How long would it take to lay 8 rows of 18 bricks each at a rate of 4 bricks per minute? Express your answer in hours. _____

5. Apples are on sale for $0.90/pound. One pound is about 4 apples. Trisha purchased a crate of apples for $10. About how many apples should the crate contain? _____

Practice

6. $7 = 2 - 5m$; $m =$ _____

7. $x + 0.054 = -1.802$; $x =$ _____

STUDY LINK 8·2 More Rate Problems and Proportions

For each of the following problems, first complete the rate table. Use the table to write an open proportion. Solve the proportion. Then write the answer to the problem.

1. A science museum requires 3 adult chaperones for every 15 students on a field trip. How many chaperones would be needed for a group of 125 students?

adults	3	a
students	15	

$$\frac{\boxed{}}{\boxed{}} = \frac{\boxed{}}{\boxed{}}$$

Answer: A group of 125 students would need _____ adult chaperones.

2. Crust and Crunch Deli sells 30 salads for every 48 sandwiches. At this rate, how many salads will they sell for every 64 sandwiches?

salads	30	w
sandwiches	48	

$$\frac{\boxed{}}{\boxed{}} = \frac{\boxed{}}{\boxed{}}$$

Answer: For every 64 sandwiches, they will sell _____ salads.

3. Tonya's car averages 240 miles for each 10 gallons of gasoline. How many gallons of gasoline will the car need to travel 216 miles?

miles	240	
gallons	10	

$$\frac{\boxed{}}{\boxed{}} = \frac{\boxed{}}{\boxed{}}$$

Answer: Tonya's car needs _____ gallons of gasoline to travel 216 miles.

4. There are 60.96 centimeters in 2 feet. How many centimeters are in 1 yard?

cm	60.96	
ft	2	

$$\frac{\boxed{}}{\boxed{}} = \frac{\boxed{}}{\boxed{}}$$

Answer: There are _____ centimeters in 1 yard.

Practice

Estimate each quotient and then divide. Round your answer to the nearest whole number. Show your work on the back of this sheet.

5. $38{,}419 \div 57$ is about _____. $38{,}419 \div 57 =$ _____

6. $7{,}648 \div 84$ is about _____. $7{,}648 \div 84 =$ _____

7. $86.5 \div 2.5$ is about _____. $86.5 \div 2.5 =$ _____

Calculating Rates

If necessary, draw a picture, find a per-unit rate, make a rate table, or use a proportion to help you solve these problems.

1. A can of worms for fishing costs $2.60. There are 20 worms in a can.

   a. What is the cost per worm? _____

   b. At this rate, how much would 26 worms cost? _____

2. An 11-ounce bag of chips costs $1.99.

   a. What is the cost per ounce, rounded to the nearest cent? _____

   b. What is the cost per pound, rounded to the nearest cent? _____

3. Just 1 gram of venom from a king cobra snake can kill 150 people. At this rate, about how many people would 1 kilogram kill? _____

4. A milking cow can produce nearly 6,000 quarts of milk each year. At this rate, about how many gallons of milk could a cow produce in 5 months? _____

5. A dog-walking service costs $2,520 for 6 months.

   What is the cost for 2 months? _____   For 3 years? _____

Try This

6. A 1-pound bag of candy containing 502 pieces costs 16.8 cents per ounce. What is the cost of 1 piece of candy? Circle the best answer.

        1.86 cents   2.99 cents   0.33 cent $\frac{1}{2}$ cent

7. Mr. Rainier's car uses about 1.6 fluid ounces of gas per minute when the engine is idling. One night, he parked his car but forgot to turn off the motor. He had just filled his tank. His tank holds 12 gallons.

   About how many hours will it take before his car runs out of gas? _____

   Explain what you did to find the answer.

   _____

   _____

   Sources: 2201 Fascinating Facts; Everything Has Its Price

STUDY LINK
8·4

Food Costs as Unit Rates

Visit a grocery store with a parent or guardian. Select 10 different items and record the cost and weight of each item in Part A of the table below.

◆ Select items that include a wide range of weights.

◆ Select only items whose containers list weights in pounds and ounces or a combination of pounds and ounces, such as 2 lb 6 oz.

◆ Do not choose produce items (fruits and vegetables).

◆ Do not choose liquids that are sold by volume (gallons, quarts, pints, liters, milliliters, or fluid ounces).

1. Complete Part A of the table at the store.

2. Complete Parts B and C of the table by

◆ converting each weight to ounces and pounds.

◆ calculating the unit cost in cents per ounce and in dollars per pound.

Example: A jar of pickles weighs 1 lb 5 oz and costs $2.39.

Convert Weight	Calculate Unit Cost
to ounces: 1 lb 5 oz = 21 oz	in cents per ounce: $\dfrac{\$2.39}{21\ oz} = \dfrac{11.4\ cents}{1\ oz}$
to pounds: 1 lb 5 oz = $1\frac{5}{16}$ lb = 1.31 lb	in dollars per pound: $\dfrac{\$2.39}{1.31\ lb} = \dfrac{\$1.82}{1\ lb}$

Part A			Part B		Part C	
Item	Cost	Weight Shown	Weight in Ounces	Cents per Ounce	Weight in Pounds	Dollars per Pound

Calculating Calories from Fat

STUDY LINK 8·5

SRB 49

1. Choose 5 breakfast items from the menu at the right. Pay no attention to total calories, but try to limit the percent of calories from fat to 30% or less. Put a check mark next to each of your 5 items.

Food	Total Calories	Calories from Fat
Toast (1 slice)	70	10
Corn flakes (8 oz)	95	trace
Oatmeal (8 oz)	130	20
Butter (1 pat)	25	25
Doughnut	205	105
Jam (1 tbs)	55	trace
Pancakes (butter, syrup)	180	60
Bacon (2 slices)	85	65
Yogurt	240	25
Sugar (1 tsp)	15	0
Scrambled eggs (2)	140	90
Fried eggs (2)	175	125
Hash browns	130	65
Skim milk (8 fl oz)	85	0
2% milk (8 fl oz)	145	45
Blueberry muffin	110	30
Orange juice (8 fl oz)	110	0
Bagel	165	20
Bagel with cream cheese	265	105

2. Record the 5 items you chose in the table. Then find the total number of calories for each column.

Food	Total Calories	Calories from Fat
Total		

3. What percent of the total number of calories comes from fat? _____

191

STUDY LINK 8·6 | **Solving Ratio Problems**

Solve the following problems. Use coins or 2-color counters to help you.
If you need to draw pictures, use the back of this page.

1. You have 45 coins. Five out of every 9 are HEADS
   and the rest are TAILS. How many coins are HEADS? _____ coins

2. You have 36 coins. The ratio of HEADS to
   TAILS is 3 to 1. How many coins are HEADS? _____ coins

3. You have 16 coins that are HEADS up and 18 coins that are TAILS up. After you
   add some coins that are TAILS up, the ratio of HEADS up to TAILS up is 1 to 1.5.

   How many coins are TAILS up? _____ coins   How many coins in all? _____ coins

4. At Richards Middle School, there are 448 students and 32 teachers.
   The San Miguel Middle School has 234 students and 18 teachers.
   Which school has a better ratio of students to
   teachers; that is, fewer students per teacher? _____

   Explain how you found your answer. _____

   _____

   _____

5. You have 6 shelves for books. Numbers
   of books are listed in the table at the right.
   The ratio of mystery books to adventure
   books to humor books is the same on
   each shelf. Complete the table.

Shelf	Mystery Books	Adventure Books	Humor Books
1	4	10	18
2	6		
3			
4		25	
5	12		
6			63

Practice

Write each quotient as a 2-place decimal.

6. 12)178   7. 84)7,428   8. 36)434

STUDY LINK 8·7   Body Composition by Weight

Solve using any method of your choice.

SRB 51 52

1. About 1 out of every 5 pounds of an average adult's
   body weight is fat. What percent of body weight is fat? _____%

2. About 60% of the human body is water. At this rate, how
   many pounds of water are in the body of a 95-pound person? _____ lb

For Problems 3–6, use a variable to represent each part, whole, or percent that is known.
Set up and solve each proportion.

3. The width of a singles tennis court is 75% the width of a doubles court.
   A doubles court is 36 ft wide. How wide is a singles court?

   $\dfrac{\text{(width of singles court)}}{\text{(width of doubles court)}}$ $\dfrac{\boxed{}}{\boxed{}} = \dfrac{\boxed{}}{\boxed{100}}$ A singles court is _____ ft wide.

4. Nadia put $500 into a savings account. At the end of 1 year, she had earned
   $30 in interest. What interest rate was the bank paying?

   $\dfrac{\text{(interest)}}{\text{(savings account)}}$ $\dfrac{\boxed{}}{\boxed{}} = \dfrac{\boxed{}}{\boxed{100}}$ The bank was paying

   _____% interest.

5. 15 is 6% of what number?

   $\dfrac{\text{(part)}}{\text{(whole)}}$ $\dfrac{\boxed{}}{\boxed{}} = \dfrac{\boxed{}}{\boxed{100}}$ 15 is 6% of _____.

6. 25% of what number is $10\frac{1}{2}$?

   $\dfrac{\text{(part)}}{\text{(whole)}}$ $\dfrac{\boxed{}}{\boxed{100}} = \dfrac{\boxed{}}{\boxed{}}$ 25% of _____ is $10\frac{1}{2}$.

Practice

7. $5\frac{2}{3} + 6\frac{3}{8} =$ _____

8. $3\frac{2}{3} - \frac{7}{9} =$ _____

9. $2\frac{3}{5} * 1\frac{3}{4} =$ _____

10. $5\frac{1}{4} \div 1\frac{3}{7} =$ _____

STUDY LINK 8·8 | **Home Data**

1. Record the following data about all the members of your household.

    a. Total number of people _____

    b. Number of males _____

    c. Number of females _____

    d. Number of left-handed people _____

    e. Number of right-handed people _____ (For people who are ambidextrous, record the hand most often used for writing.)

For the rectangles in this Study Link, use length as the measure of the longer sides and width as the measure of the shorter sides.

2. Find an American flag or a picture of one. Measure its length and width.

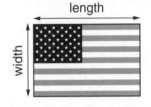

    a. length _____ (unit)

    b. width _____ (unit)

3. Measure the length and width of a television screen to the nearest $\frac{1}{2}$ inch.

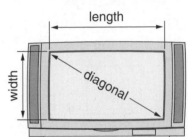

    a. length _____ (unit)

    b. width _____ (unit)

4. Find 3 books of different sizes, such as a small paperback, your math journal, and a large reference book. Measure the length and width of each book to the nearest $\frac{1}{2}$ inch.

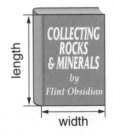

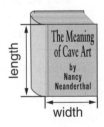

    a. Small book: length _____ (unit) width _____ (unit)

    b. Medium book: length _____ (unit) width _____ (unit)

    c. Large book: length _____ (unit) width _____ (unit)

197

STUDY LINK
8·8

Home Data *continued*

5. Find samples of the following items. Measure the length and width of each to the nearest $\frac{1}{4}$ inch.

   a. Postcard

   length _____ (unit)   width _____ (unit)

   b. Index card

   length _____ (unit)   width _____ (unit)

   c. Envelope (regular)

   length _____ (unit)   width _____ (unit)

   d. Envelope (business)

   length _____ (unit)   width _____ (unit)

   e. Notebook paper

   length _____ (unit)   width _____ (unit)

6. Show the 4 rectangles below to each member of your household. Ask each person to select the rectangle that he or she likes best or thinks is the nicest looking. Tally the answers. Remember to include your own choice.

A
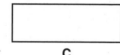
B

C
D

Voting Results	A	B	C	D
Tally of Votes				
Number of Votes				

7. Measure the rise and run of stairs in your home. The diagram shows what these dimensions are. (If there are no stairs in your home, measure stairs outdoors or in a friend's or neighbor's home.)

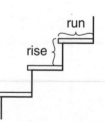

   a. rise _____ in.   **b.** run _____ in.

Practice

Round each quotient to the nearest tenth.

8. $32\overline{)74.9}$ = _____   **9.** $15\overline{)864.9}$ = _____   **10.** $68\overline{)696.1}$ = _____

STUDY LINK
8·9
Scale Drawings

SRB
121 122

Measure the object in each drawing to the nearest millimeter. Then use the size-change factor to determine the actual size of the object.

Size-change Factor: $\dfrac{\text{changed length}}{\text{actual length}}$

1. **a.** Diameter in drawing: _____

  b. Actual diameter: _____

button

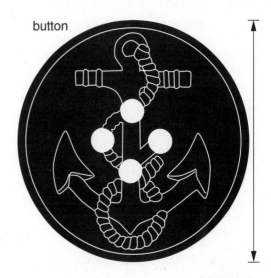

Size Change	Size-change Factor
Scale 2:1	

2. **a.** Height in drawing:

  _____

  b. Actual height:

  _____

glue bottle

CRAFT GLUE

Size Change	Size-change Factor
$\frac{1}{4}$X	

3. **a.** Length in drawing:

  _____

  b. Actual length:

  _____

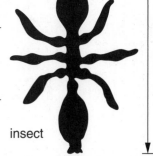

insect

Size Change	Size-change Factor
Scale 3:1	

4. **a.** Height in drawing: _____

  b. Actual height: _____

Size Change	Size-change Factor
Scale 1:3	

watering can

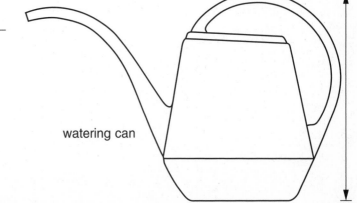

Similar Polygons

1. Triangles *JKL* and *RST* are similar.

   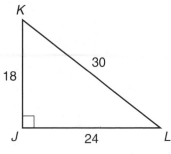

   a. Find the ratio *KL:ST*. _____

   b. m∠*R* = _____

   c. The length of $\overline{RS}$ = _____

   d. $\dfrac{\text{perimeter of } \triangle JKL}{\text{perimeter of } \triangle RST}$ = _____

2. Quadrangles *ABCD* and *MLON* are similar.

   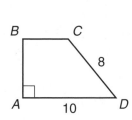 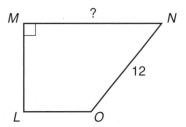

   a. The length of $\overline{MN}$ = _____

   b. The size-change factor: $\dfrac{\text{large trapezoid}}{\text{small trapezoid}}$ = _____ X

3. Find the distance (*d*) across the pond if the small triangle is similar to the large triangle.

   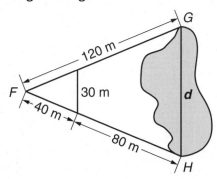

   d = _____ m

Practice

Round each number to the nearest thousandth.

4. 0.00673 _____

5. 63.4982 _____

6. 4.8919 _____

7. 5.9198 _____

STUDY LINK 8·11 **Comparing Ratios**

SRB
117–119

1. A dictionary measures 24 centimeters by 20 centimeters.

   The ratio of its length to its width is about _____ to 1.

   Explain. _____

2. A sheet of legal-size paper measures 14 inches by
   $8\frac{1}{2}$ inches. The ratio of its length to its width is about _____ to 1.

   Is this the same ratio as for a sheet of paper that measures 11 inches by $8\frac{1}{2}$ inches? _____

   Explain. _____

   _____

3. Jeffrey answered 28 out of 30 problems correctly on his math test.
   Lucille answered 47 out of 50 problems correctly on her math test.
   Who did better on the test? _____

   Explain. _____

   _____

4. A ruler is 30 centimeters long and 2.5 centimeters
   wide. The ratio of its length to its width is about _____ to 1.

Try This

5. If a ruler is 33.6 centimeters long, how wide would it have to be to
   have the same length-to-width ratio as the ruler in Problem 4? _____ centimeters

   Explain. _____

   _____

Practice

6.   74   7.   366   8.   483   9.   806
   * 12   * 58   * 32   * 157

203

STUDY LINK 8·12 | **Rate and Ratio Review**

1. Match each ratio on the left with one of the ratios on the right.

   a. Circumference to diameter of a circle _____ 1.6 to 1

   b. Length to width of a Golden Rectangle _____ 3 to 5

   c. Diameter to radius of a circle _____ 2 to 1

   d. Length of one side of a square to another _____ 3.14 to 1

   e. 12 correct answers out of 20 problems _____ 1 to 1

2. Refer to the following numbers to answer the questions below.

   1   2   3   4   5   6   7   8   9   10

   a. What percent of the numbers are prime numbers? _____

   b. What is the ratio of numbers divisible
      by 3 to numbers divisible by 2?   _____

3. A 12-pack of Chummy Cola costs $3 at Stellar Supermart.

   a. Complete the rate table at the right
      to find the per-unit rates.

dollars		3.00	1.00
cans	1	12	

   b. At this price, how much would 30 cans of Chummy Cola cost? _____

   c. How many cans could you buy for $2.00? _____

4. Complete or write a proportion for each problem. Then solve the problem.

   a. Only $\frac{4}{9}$ of the club members voted in the last election. There are
      54 members in the club. How many members voted?

      Proportion $\frac{4}{9} = \frac{x}{54}$ Answer _____

   b. During basketball practice, Christina made 3 out of every 5 free throws she
      attempted. If she made 12 free throws, how many free throws did she attempt in all?

      Proportion _____   Answer _____

205

STUDY LINK 8·12   Rate and Ratio Review *continued*

117–122

5. a. Draw circles and squares so the ratio of circles to squares is 3 to 2 and the total number of shapes is 10.

b. Draw circles and squares so the ratio of circles to total shapes is 2 to 3 and the total number of squares is 2.

c. Draw circles and squares so the ratio of circles to squares is 1 to 3 and the total number of shapes is 12.

6. The city is planning to build a new park. The park will be rectangular in shape, approximately 800 feet long and 625 feet wide. Make a scale drawing of the park on the $\frac{1}{2}$-inch grid paper below.

Scale: $\frac{1}{2}$ inch represents 100 feet.

STUDY LINK 8·13 | **Unit 9: Family Letter**

More about Variables, Formulas, and Graphs

You may be surprised at some of the topics that are covered in Unit 9. Several of them would be traditionally introduced in a first-year algebra course. If you are assisting your child, you might find it useful to refer to the *Student Reference Book* to refresh your memory.

Your child has been applying many mathematical properties, starting as early as first grade. In Unit 9, the class will explore and apply one of these properties, the distributive property, which can be stated as follows:

For any numbers *a, b,* and *c,* $a * (b + c) = (a * b) + (a * c)$.

Students will use this property to simplify algebraic expressions. They will use these procedures, together with the equation-solving methods that were presented in Unit 6, to solve more difficult equations that contain parentheses or like terms on at least one side of the equal sign. Here is an example:

To solve the equation $5(b + 3) - 3b + 5 = 4(b - 1)$,

1. Use the distributive property to remove the parentheses. $\qquad 5b + 15 - 3b + 5 = 4b - 4$

2. Combine like terms. $\qquad\qquad\qquad\qquad 2b + 20 = 4b - 4$

3. Solve the equation. $\qquad\qquad\qquad\qquad\quad 20 = 2b - 4$

$$24 = 2b$$

$$b = 12$$

Much of Unit 9 also focuses on applying formulas—in computer spreadsheets and in calculating the areas of circles, rectangles, triangles, and parallelograms, the perimeters of polygons, and the circumferences of circles. Students will also use formulas for calculating the volumes of rectangular prisms, cylinders, and spheres to solve a variety of interesting problems.

Finally, your child will be introduced to the Pythagorean theorem, which states that if *a* and *b* are the lengths of the legs of a right triangle and *c* is the length of the hypotenuse, then $a^2 + b^2 = c^2$. By applying this theorem, students will learn how to calculate long distances indirectly—that is, without actually measuring them.

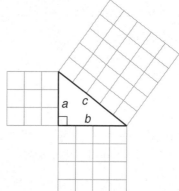

Please keep this Family Letter for reference as your child works through Unit 9.

Vocabulary

Important terms in Unit 9:

combine like terms To rewrite the sum or difference of *like terms* as a single term. For example, $5a + 6a$ can be rewritten as $11a$, because $5a + 6a = (5 + 6)a = 11a$. Similarly, $16t - 3t = 13t$.

Distributive Property of Multiplication over Addition A property relating multiplication to a sum of numbers by distributing a factor over the terms in the sum.
For example, $2 * (5 + 3) = (2 * 5) + (2 * 3) = 10 + 6 = 16$.
In symbols: For any numbers a, b, and c:
$$a * (b + c) = (a * b) + (a * c),$$
$$\text{or } a(b + c) = ab + ac$$

Distributive Property of Multiplication over Subtraction A property relating multiplication to a difference of numbers by distributing a factor over the terms in the difference.
For example, $2 * (5 - 3) = (2 * 5) - (2 * 3) = 10 - 6 = 4$.
In symbols: For any numbers a, b, and c:
$$a * (b - c) = (a * b) - (a * c),$$
$$\text{or } a(b - c) = ab - ac$$

equivalent fractions Fractions with different denominators that name the same number.

hypotenuse In a right triangle, the side opposite the right angle.

indirect measurement The determination of heights, distances, and other quantities that cannot be directly measured.

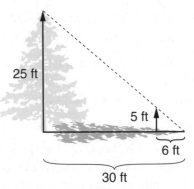

Indirect measurement lets you calculate the height of the tree from the other measure.

leg of a right triangle Either side of the right angle in a right triangle; a side that is not the *hypotenuse*.

like terms In an algebraic expression, either the constant terms or any terms that contain the same variable(s) raised to the same power(s). For example, $4y$ and $7y$ are like terms in the expression $4y + 7y - z$.

Pythagorean theorem If the *legs of a right triangle* have lengths a and b and the *hypotenuse* has length c, then $a^2 + b^2 = c^2$.

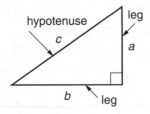

simplify an expression To rewrite an expression by clearing grouping symbols and combining *like terms* and constants.

Do-Anytime Activities

To work with your child on the concepts taught in this unit and previous units, try these interesting and rewarding activities:

1. To practice simplifying expressions and solving equations, ask your child to bring home the game materials for *Algebra Election*. Game directions are in the *Student Reference Book*.

2. If you have any mobiles in your home, ask your child to explain to you how to perfectly balance one. Have your child show you the equations he or she used to balance it.

3. Your child may need extra practice with the partial-quotients division algorithm. Have him or her show you this method. Provide a few problems to practice at home, and have your child explain the steps to you while working through them.

As You Help Your Child with Homework

As your child brings assignments home, you may want to go over the instructions together, clarifying them as necessary. The answers listed below will guide you through some of the Unit 9 Study Links.

Study Link 9·1

1. **a.** $(8 * 4) + (7 * 4)$ **b.** $(8 * 6) + (5 * 6)$
   $4 * (7 + 8)$ $6 * (5 + 8)$
   c. $(4 + 9) * 3$ $(8 + 5) * 6$
   $(9 * 3) + (4 * 3)$

2. **a.** 6
   b. $(9 - 3) * 5 = 30$ $(9 * 5) - (3 * 5) = 30$

3. **a.** N   **b.** O   **c.** O
   d. N   **e.** P   **f.** O

4. 3.92 $(8 * 0.10) + (8 * 0.39) = 3.92$

Study Link 9·2

1. **a.** $(7 * 3) + (7 * 4)$
   b. $(7 * 3) + (7 * \pi)$
   c. $(7 * 3) + (7 * y)$
   d. $(7 * 3) + (7 * (2 * 4))$
   e. $(7 * 3) + (7 * (2 * \pi))$
   f. $(7 * 3) + (7 * (2 * y))$

2. **b.** $(20 * 42) - (20 * 19) = 840 - 380 = 460$
   c. $(32 * 40) + (50 * 40) = 1,280 + 2,000 = 3,280$
   d. $(90 * 11) - (8 * 11) = 990 - 88 = 902$
   e. $(9 * 15) + (9 * 25) = 135 + 225 = 360$

3. **a.** $(80 * 5) + (120 * 5) = (80 + 120) * 5$
   c. $12(d - t) = 12d - 12t$
   d. $(a + c) * n = (a * n) + (c * n)$
   f. $(9 * \frac{1}{2}) - (\frac{1}{3} * \frac{1}{2}) = (9 - \frac{1}{3}) * \frac{1}{2}$
4. 3   5. $\frac{11}{14}$ 6. $\frac{8}{57}$

Study Link 9·3

1. $15x$ **2.** $\frac{3}{10}y$ **3.** $-11t$ **4.** d

5. -6 **6.** $3p$ **7.** -3 **8.** 8.3

9. $7b + 14$ **10.** $1\frac{1}{6}a + \frac{1}{4}t$

11. -53 **12.** 23 **13.** 132 **14.** -19

Study Link 9·4

1. $45f + 109$ **2.** $12m$ **3.** $32k + 44$

4. $-y + 2b + 24$

5. $65,800$ **6.** 0.2348 **7.** 0.5163 **8.** 0.0796

Study Link 9·5

Column 1   **Column 2**

A. $4x - 2 = 6$ C $6j + 8 = 8 + 6j$
Solution: $x = 2$ A $2c - 1 = 3$
                     B $6w = -12$
                     C $\frac{2h}{2h} = 1$
                     A $\frac{3q}{3} - 6 = -4$
                     A $3(r + 4) = 18$

B. $3s = -6$ C $2(5x + 1) = 10x + 2$
Solution: $s = -2$ A $-5x - 5(2 - x) = 2(x - 7)$
                      D $s = 0$
                      B $5b - 3 - 2b = 6b + 3$

C. $3y - 2y = y$ B $\frac{t}{4} + 3 = 2\frac{1}{2}$
Solution: $y = $ A $6z = 12$
any number   D $2a = (4 + 7)a$

D. $5a = 7a$
Solution: $a = 0$

1. 2^5 **2.** 10^2 **3.** 5^4 **4.** 4^1

Study Link 9·6

1. 7 **2.** 38 **3.** 4 **4.** 2

5. $23 + 14y$ **6.** $-2b + 32$

7. $3f - 55 - 10k$ **8.** $225 + 35g$

9. $r + 23$ **10.** $4b + 72; 72 - (-4b)$

11. $W = 5b; D = 4; w = 30; d = 12$

Equation: $5b * 4 = 30 * 12$; Solution: $b = 18$
Weight of the object on the left: 90

12. $5\frac{11}{24}$ **13.** 92 **14.** $5\frac{5}{7}$

Study Link 9·7

3. 2.7 feet

Study Link 9·8

1. 112 in.^{2} **2.** 2.5 ft^2 **3.** 108 cm^2

4. 45.5 mm^2 **5.** 55 ft^2 **6.** 696 m^2

7. $a * b$ **8.** $(n + m) * y$

9. 63.6 **10.** 0.1

Study Link 9·9

1. 120 in.^{3} **2.** 904.32 in.^{3} **3.** 11.97 in.^{2}

4. 10.4 m^3 **5.** $3,391$ yd^2 **6.** 3.22 ft^3

7. 95 **8.** 37.8 **9.** $1,400$

Study Link 9·10

1. Answers vary. **2.** Answers vary.

3. 13.48 **4.** 17.62

Study Link 9·11

1. a. $C = \frac{5}{9} * (77 - 32)$; 25°C

   b. $50 = \frac{5}{9} * (F - 32)$; 122°F

2. a. $A = \frac{1}{2} * 17 * 5$; 42.5 cm²

   b. $90 = \frac{1}{2} * 12 * h$; 15 in.

3. a. $V = \frac{1}{3} * \pi * 4 * 9$; 37.68 in.³

   b. $94.2 = \frac{1}{3} * \pi * 9 * h$; 10 cm

Study Link 9·12

1. 12   **2.** 200   **3.** 30   **4.** 0.4

5. $\frac{5}{11}$ **6.** 100   **7.** 3.46   **8.** 7.14

9. 7.94   **10.** 25 m   **11.** 9.8 ft   **12.** 22 yd

13. 127.3 ft

14. 18   **15.** 23

Study Link 9·13

1. a. $7x$ **b.** $4x + 7$ **c.** $6x + 2$ **d.** 6

2. Sample answer: Liani did not multiply 10 by 8. The simplified expression should be $8x + 80$.

3. a. $x = -10$ **b.** $g = -5$

   c. $y = 4$ **d.** $x = 14$

4. Length of $\overline{AB}$: 5 in.; Length of $\overline{BC}$: 8 in.; Length of $\overline{AC}$: 5 in.

5. 6 cm²

6. 4 blocks

7. 1.5   **8.** 1.75   **9.** 0.6

STUDY LINK 9·1 | **Multiplying Sums**

1. For each expression in the top row, find one or more equivalent expressions below it. Fill in the oval next to each equivalent expression.

a. $(8 + 7) * 4$

   O $(8 * 4) + (7 * 4)$

   O $4 * (7 + 8)$

   O $(8 + 4) * 7$

   O $(8 + 4) * (7 + 4)$

b. $(6 * 5) + (6 * 8)$

   O $(8 * 6) + (5 * 6)$

   O $6 * (5 + 8)$

   O $(8 + 5) * 6$

   O $(6 + 5) * (6 + 8)$

c. $3 * (9 + 4)$

   O $(9 + 4) * (3 + 4)$

   O $9 * (3 + 4)$

   O $(4 + 9) * 3$

   O $(9 * 3) + (4 * 3)$

2. The area of Rectangle M is 45 square units.

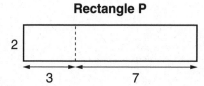

Rectangle M

a. What is the value of b? _____

b. Write 2 different number sentences to describe the area of the unshaded part of Rectangle M.

(____ − ____) * ____ = ____   (____ * ____) − (____ * ____) = ____

3. Each of the following expressions describes the area of one of the rectangles below. Write the letter of the rectangle next to its expression.

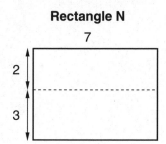

Rectangle N

Rectangle O

Rectangle P

a. $(3 + 2) * 7$ _____

b. $(2 * 3) + (7 * 3)$ _____

c. $(7 + 2) * 3$ _____

d. $(3 * 7) + (2 * 7)$ _____

e. $2 * (7 + 3)$ _____

f. $3 * (2 + 7)$ _____

4. Sandra wants to buy envelopes and stamps to send cards to 8 friends. Envelopes cost $0.10 and stamps cost $0.39. How much will she spend? _____

Write a number model to show how you solved the problem.

STUDY LINK 9·2 | **Using the Distributive Property**

> **Reminder:** $a * (x + y) = (a * x) + (a * y)$
> $a * (x - y) = (a * x) - (a * y)$

SRB
248 249

1. Use the distributive property to rewrite each expression.

  a. $7 * (3 + 4) = ($_____ $*$ _____$) + ($_____ $*$ _____$)$

  b. $7 * (3 + \pi) = ($_____ $*$ _____$) + ($_____ $*$ _____$)$

  c. $7 * (3 + y) = ($_____ $*$ _____$) + ($_____ $*$ _____$)$

  d. $7 * (3 + (2 * 4)) = ($_____ $*$ _____$) + ($_____ $* (2 * 4))$

  e. $7 * (3 + (2 * \pi)) = ($_____ $*$ _____$) + ($_____ $* (2 *$ _____$))$

  f. $7 * (3 + (2 * y)) = ($_____ $*$ _____$) + ($_____ $* ($_____ $*$ _____$))$

2. Use the distributive property to solve each problem. Study the first one.

  a. $7 * (110 + 25) =$ *(7 * 110) + (7 * 25) = 770 + 175 = 945*

  b. $20 * (42 - 19) =$ _____

  c. $(32 + 50) * 40 =$ _____

  d. $(90 - 8) * 11 =$ _____

  e. $9 * (15 + 25) =$ _____

3. Circle the statements that are examples of the distributive property.

  a. $(80 * 5) + (120 * 5) = (80 + 120) * 5$ **b.** $6 * (3 - 0.5) = (6 * 3) - 0.5$

  c. $12(d - t) = 12d - 12t$ **d.** $(a + c) * n = a * n + c * n$

  e. $(16 + 4m) + 9.7 = 16 + (4m + 9.7)$ **f.** $(9 * \frac{1}{2}) - (\frac{1}{3} * \frac{1}{2}) = (9 - \frac{1}{3}) * \frac{1}{2}$

Practice

Write each quotient in lowest terms.

4. $\frac{1}{5} \div \frac{1}{15}$ _____

5. $\frac{3}{7} \div \frac{6}{11}$ _____

6. $1\frac{1}{19} \div 7\frac{1}{2}$ _____

215

STUDY LINK 9·3 Combining Like Terms

Simplify each expression by rewriting it as a single term.

1. $3x + 12x =$ _____

2. $(1\frac{3}{5})y - (1\frac{3}{10})y =$ _____

3. $-(5t) - 6t =$ _____

4. $4d + (-3d) =$ _____

Complete each equation.

5. $15k = (9 -$ _____$)k$

6. $3.6p - p =$ _____ $- 0.4p$

7. $(8 +$ _____$) * m = 5m$

8. _____$j - 4.5j = 3.8j$

Simplify each expression by combining like terms. Check your answers by substituting the given values for the variables. Show your work on the back of this sheet.

Example: $18 + 6m + 2m + 26$

Combine the m terms. $6m + 2m = 8m$

Combine the number, or constant, terms. $18 + 26 = 44$

So, $18 + 6m + 2m + 26 = 8m + 44$.

Check: Substitute 5 for m.

$18 + (6 * 5) + (2 * 5) + 26 = (8 * 5) + 44$

$18 + 30 + 10 + 26 = 40 + 44$

$84 = 84$

9. $8b + 9 + 4b - 3b + (-2b) - (-5) =$ _____

Check for: $b = -6$

10. $\frac{1}{2}a + \frac{3}{4}t + \frac{2}{3}a + (-\frac{1}{2}t) =$ _____

Check for: $a = 2$ and $t = -2$

Practice

11. $-117 + 64 =$ _____

12. $-9 - (-32) =$ _____

13. $-12 * (-11) =$ _____

14. $\frac{57}{-3} =$ _____

STUDY LINK
9·4

Simplifying Expressions

Simplify each expression by removing parentheses and combining like terms. Check by substituting the given values for the variables. Show your work.

1. $7(7 + 5f) + (f + 6)10$ _____

Check: Substitute $-\frac{1}{5}$ for f.

2. $3(4 + 5m) - 12 + (-3m)$ _____

Check: Substitute $\frac{1}{3}$ for m.

3. $(12 - 3 + 5k)6 + 4k - 2(k + 5)$ _____

Check: Substitute 0.5 for k.

4. $5(y - b) + 3b - 6y + 4(6 + b)$ _____

Check: Substitute 1 for y and $\frac{2}{3}$ for b.

Practice

Find each product or quotient.

5. $0.658 * 10^5$ _____

6. $234.8 \div 10^3$ _____

7. $5,163 * 10^{-4}$ _____

8. $7.96 \div 10^2$ _____

STUDY LINK 9·5   **Equivalent Equations**

Each equation in Column 2 is equivalent to an equation in Column 1.
Solve each equation in Column 1. Write *Any number* if all numbers are
solutions of the equation.

Match each equation in Column 1 with an equivalent equation in Column 2.
Write the letter label of the equation in Column 1 next to the equivalent
equation in Column 2.

Column 1	**Column 2**

Column 1

A $4x - 2 = 6$

Solution _____

B $3s = -6$

Solution _____

C $3y - 2y = y$

Solution _____

D $5a = 7a$

Solution _____

Column 2

_____ $6j + 8 = 8 + 6j$

_____ $2c - 1 = 3$

_____ $6w = -12$

_____ $\frac{2h}{2h} = 1$

_____ $\frac{3q}{3} - 6 = -4$

___*A*___ $3(r + 4) = 18$

_____ $2(5x + 1) = 10x + 2$

_____ $-5x - 5(2 - x) = 2(x - 7)$

_____ $s = 0$

_____ $5b - 3 - 2b = 6b + 3$

_____ $\frac{t}{4} + 3 = 2\frac{1}{2}$

_____ $6z = 12$

_____ $2a = (4 + 7)a$

Practice

Write each product or quotient in exponential notation.

1. $2^2 * 2^3$ _____   **2.** $\frac{10^4}{10^2}$ _____   **3.** $5^2 * 5^2$ _____   **4.** $\frac{4^3}{4^2}$ _____

STUDY LINK 9·6 Expressions and Equations

Solve.

1. $3x + 9 = 30$ $x =$ _____

2. $73 = \frac{1}{2}(108 + f)$ $f =$ _____

3. $55 = (9 - d) * 11$ $d =$ _____

4. $(m * 15) + (m * 6) = 42$ $m =$ _____

Simplify these expressions by combining like terms.

5. $8y + 27 + 6y + (-4)$ _____

6. $7b + 17 - 9b + 15$ _____

7. $3f - 80 + 25 - 10k$ _____

8. $240 + 5g + 3(10g - 5)$ _____

Circle all expressions that are equivalent to the original. There may be more than one. Check your answer by substituting values for the variable.

9. Original: $3r + 17 - 2r + 6$

   $5r + 23$ $23 - r$ $r + 23$ $13 + r$

10. Original: $8(9 + b) - 4b$

    $89 - 3b$ $72 - 3b$ $4b + 72$ $72 - (-4b)$

Try This

11. The top mobile is in balance. The fulcrum is at the center of the rod. A mobile will balance when $W * D = w * d$.

    Look at the bottom mobile. What is the weight of the object on the left?

    Write and solve an equation to answer the question.

    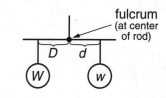

    $W =$ _____ $D =$ _____ $w =$ _____ $d =$ _____

    Equation _____   Solution _____

    The weight of the object on the left is _____ units.

Practice

12. $8\frac{1}{3} - 2\frac{7}{8}$ _____

13. $3\frac{5}{6} * 24$ _____

14. $25 \div 4\frac{3}{8}$ _____

STUDY LINK 9·7 **Circumferences and Areas of Circles**

Circles			⊠
	A	B	C
1	circumferences and areas of circles		
2	radius (ft)	circumference (ft)	area (ft²)
3	r	$2\pi r$	πr^2
4	0.5		
5	1.0		
6	1.5	9.4	7.1
7	2.0	12.6	12.6
8	2.5		
9	3.0		

1. Complete the spreadsheet at the left. For each radius, calculate the circumference and area of a circle having that radius. Round your answers to tenths.

SRB
140
213 218

2. Use the data in the spreadsheet to graph the number pairs for radius and circumference on the first grid below. Then graph the number pairs for radius and area on the second grid below. Connect the plotted points.

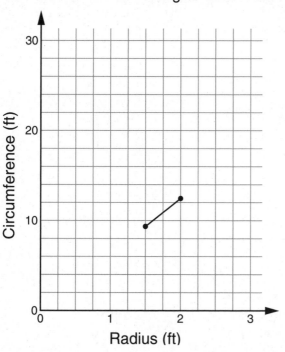

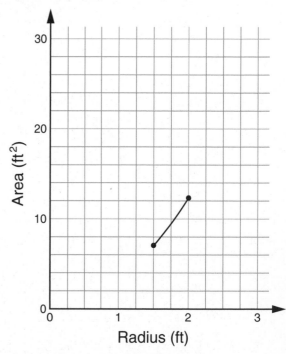

3. A circular tabletop has an area of 23 square feet. Use the second line graph to estimate the radius of the tabletop.   Radius: About _____

(unit)

225

STUDY LINK
9·8 **Area Problems**

Calculate the area of each figure in Problems 1–6. Remember to include
the unit in each answer.

SRB
215–217

1. parallelogram

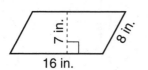

7 in. 8 in.
16 in.

Area _____

2. rectangle

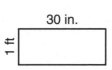

30 in.
1 ft

Area _____

3. parallelogram

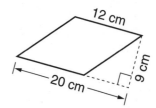

12 cm
20 cm
9 cm

Area _____

4. triangle

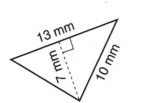

13 mm
7 mm
10 mm

Area _____

5. triangle

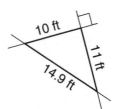

10 ft
11 ft
14.9 ft

Area _____

6. trapezoid

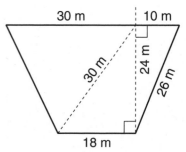

30 m   10 m
30 m   24 m   26 m
18 m

Area _____

Try This

In Problems 7 and 8, all dimensions are given as variables. Write a true
statement in terms of the variables to express the area of each figure.

Example:

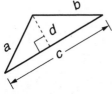

b
a   d
c

Area $\frac{1}{2} * c * d$

7.

c
b
a   a

Area _____

8.

x   y
m   n

Area _____

Practice

9. $x \div 5.3 = 12$ $x =$ _____

10. $-3.1 = -31w$ $w =$ _____

Area formulas

Rectangle:	$A = b * h$
Parallelogram:	$A = b * h$
Triangle:	$A = \frac{1}{2} * b * h$

A	= area
V	= volume
B	= area of base
C	= circumference
b	= length of base
h	= height
l	= length
w	= width
r	= radius

Volume formulas

Cylinder:	$V = B * h = \pi * r^2 * h$
Rectangular prism:	$V = B * h = l * w * h$
Sphere:	$V = \frac{4}{3} * \pi * r^3$

Circumference formula $C = 2\pi r$

Calculate the area or volume of each figure. Pay close attention to the units.

1.

6"
4"
5"

Volume _____
(unit)

2.

diameter = 12"
Use 3.14 for π.

Volume _____
(unit)

3.

5.7"
2.1"

Area _____
(unit)

4.

1.6 m
Use 3.14 for π.
6.5 m²

Volume _____
(unit)

5.

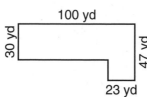

100 yd
30 yd
47 yd
23 yd

Area _____
(unit)

Try This

6.

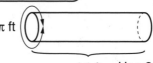

π ft

4.1 ft   Use 3.14 for π.

Volume _____
(unit)

Practice

7. 0.95 m = _____ cm   **8.** 378 mm = _____ cm   **9.** 1.4 m = _____ mm

229

STUDY LINK
9·10

Solving Equations by Trial and Error

Find numbers that are close to the solution of each equation.
Use the suggested test numbers to get started.

SRB
241–243

1. Equation: $r^2 + r = 15$

r	r^2	$r^2 + r$	Compare $r^2 + r$ to 15.
3	9	12	< 15
4	16	20	> 15
3.5	12.25	15.75	> 15

My closest solution _____

2. Equation: $x^2 - 2x = 23$

x	x^2	$2x$	$x^2 - 2x$	Compare $x^2 - 2x$ to 23.
6	36	12	24	> 23
5	25	10	15	< 23
5.5	30.25	11	19.25	< 23

My closest solution _____

Practice

3. $56 - 42.52 =$ _____

4. $23.5 - 5.88 =$ _____

STUDY LINK 9·11 **Using Formulas**

Each problem below states a formula and gives the values of all but one of the variables in the formula. Substitute the known values for the variables in the formula and then solve the equation.

SRB 245 246

1. The formula $C = \frac{5}{9} * (F - 32)$ may be used to convert between Fahrenheit and Celsius temperatures.

   a. Convert 77°F to degrees C.

   Equation _____
   Solve.

   77°F = _____ °C

   b. Convert 50°C to degrees F.

   Equation _____
   Solve.

   50°C = _____ °F

2. The formula for the area of a trapezoid is $A = \frac{1}{2} * (a + b) * h$.

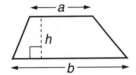

   a. Find the area (A) of a trapezoid if $a = 7$ cm, $b = 10$ cm, and $h = 5$ cm.

   Equation _____
   Solve.

   Area _____
         (unit)

   b. Find the height (h) of a trapezoid if $a = 6.5$ inches, $b = 5.5$ inches, and $A = 90$ inches^2.

   Equation _____
   Solve.

   Height _____
         (unit)

3. The formula for the volume of a cone is $V = \frac{1}{3} * \pi * r^2 * h$.
   Use 3.14 for π.

   a. Find the volume (V) of a cone if $r = 2$ inches and $h = 9$ inches.

   Equation _____
   Solve.

   Volume _____
         (unit)

   b. Find the height (h) of a cone if $r = 3$ cm and $V = 94.2$ cm^3.

   Equation _____
   Solve.

   Height _____
         (unit)

233

STUDY LINK
9·12 Pythagorean Theorem

SRB
167
285 286

Mentally find the positive square root of each number.

1. $\sqrt{144} =$ _____

2. $\sqrt{200^2} =$ _____

3. $\sqrt{900} =$ _____

4. $\sqrt{0.16} =$ _____

5. $\sqrt{\dfrac{25}{121}} =$ _____

6. $\sqrt{10,000} =$ _____

Use a calculator to find each square root. Round to the nearest hundredth.

7. $\sqrt{12} =$ _____

8. $\sqrt{51} =$ _____

9. $\sqrt{63} =$ _____

Use the Pythagorean theorem to find each missing length. Round your answer to the nearest tenth.

10.

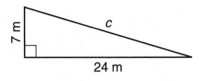

7 m
c
24 m

$c =$ _____
(unit)

11.

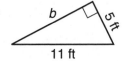

b
5 ft
11 ft

$b =$ _____
(unit)

12.

122 yd
120 yd
a

$a =$ _____
(unit)

13. Find the distance (d) from home plate to second base.

$d =$ _____ ft

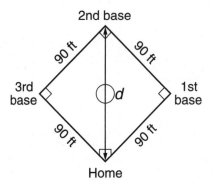

2nd base
90 ft
90 ft
3rd base
1st base
d
90 ft
90 ft
Home

Practice

Simplify.

14. $2[9(6 - 5)] =$ _____

15. $5 + 3 * 4 - 8 + 2 * 7 =$ _____

STUDY LINK 9·13 | **Unit 9 Review**

SRB 251 252

1. Simplify the following expressions by combining like terms.

a. $4x + 3x =$ _____

b. $3x + 7 + x =$ _____

c. $4 * (x + 2) + 2x - 6 =$ _____

d. $(x + 3) * 2 - 2x =$ _____

2. Liani simplified the expression $8(x + 10)$ as $(8 * x) + 10$. What did she do wrong? Explain her mistake and show the correct way to solve the problem.

_____

_____

3. Solve each equation. Show your work on the back of this sheet.

a. $3x - 4 = 4x + 6$ _____

b. $5 * (2 - 6) = 4g$ _____

c. $3(2y - 3) = 15$ _____

d. $\frac{(2x - 1)}{3} = 9$ _____

4. The perimeter of triangle *ABC* is 18 inches. What is the length of each side?

AB _____ BC _____ AC _____

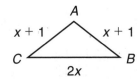

5. The perimeter of right triangle *GLD* is 12 centimeters.

What is the area of the triangle? _____

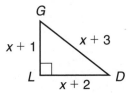

6. Toshi often walks to school along Main Street and Elm Street. If he were to take Pythagoras Avenue instead, how many fewer blocks would he walk? _____

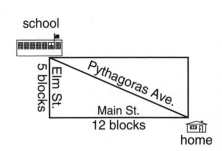

Practice

7. $28\overline{)42} =$ _____

8. $161 \div 92 =$ _____

9. $200\overline{)120} =$ _____

Unit 10: Family Letter

Geometry Topics

Unit 10 includes a variety of activities involving some of the more recreational, artistic, and lesser-known aspects of geometry. In *Fifth Grade Everyday Mathematics*, students explored same-tile **tessellations.** A tessellation is an arrangement of closed shapes that covers a surface completely, without gaps or overlaps. Your kitchen or bathroom floor may be an example of a tessellation. A regular tessellation involves only one kind of regular polygon. Three examples are shown at the right.

In Unit 10 of *Sixth Grade Everyday Mathematics*, your child will explore semiregular tessellations. A **semiregular tessellation** is made from two or more kinds of regular polygons. For example, a semiregular tessellation can be made from equilateral triangles and squares as shown below.

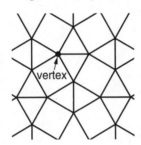

vertex

The angles around every vertex point in a semiregular tessellation must be congruent to the angles around every other vertex point. Notice that at each vertex point in the tessellation above, there are the vertices of three equilateral triangles and two squares, always in the same order.

The artist M. C. Escher used **transformation geometry**—translations, reflections, and rotations of geometric figures—to create intriguing tessellation art. Ask your child to show you the translation tessellation that students created in the style of Escher.

Your child will also explore topology. **Topology**, sometimes called *rubber-sheet geometry,* is a modern branch of geometry that deals with, among other topics, properties of geometric objects that do not change when the objects' shapes are changed. Ask your child to share with you some ideas from topology, such as Möbius strips.

Please keep this Family Letter for reference as your child works through Unit 10.

Math Tools

Your child will use the **Geometry Template** to explore and design tessellations. This tool includes a greater variety of shapes than the pattern-block template from previous grades. It might more specifically be called a geometry-and-measurement template. The measuring devices include inch and centimeter scales, a Percent Circle useful for making circle graphs, and a full-circle and a half-circle protractor.

Vocabulary

Important terms in Unit 10:

genus In *topology,* the number of holes in a geometric shape. Shapes with the same genus are topologically equivalent. For example, a doughnut and a coffee cup are equivalent because both are genus 1.

Genus 0   Genus 1

Möbius strip (Möbius band) A 3-dimensional figure with only one side and one edge, named for the German mathematician August Ferdinand Möbius (1790–1868).

Möbius strip

order of rotation symmetry The number of times a rotation image of a figure coincides with the figure before completing a 360° rotation.

A figure with order 5
rotation symmetry

regular polygon A polygon in which all sides are the same length and all angles have the same measure.

Regular polygons

regular tessellation A tessellation of one *regular polygon.* The three regular tessellations are shown below.

The three regular tessellations

rotation symmetry A figure has rotation symmetry if it is the rotation image of itself after less than a full turn around a center or axis of rotation.

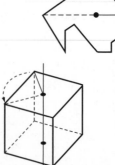

Shapes with rotation symmetry

topological transformation A transformation that pairs a figure with its image after shrinking, stretching, twisting, bending, or turning inside out. Tearing, breaking, and sticking together are not allowed. Shapes that can be changed into one another by a topological transformation are called "topologically equivalent shapes." For example, a doughnut is topologically equivalent to a coffee cup.

translation tessellation A *tessellation* made of a tile in which one or more sides is a translation image of the opposite side(s). Dutch artist M. C. Escher (1898–1972) created many beautiful and elaborate translation tessellations.

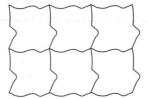

A translation tessellation

vertex point A point where the corners of tessellation tiles meet.

Do-Anytime Activities

To work with your child on the concepts taught in this unit, try these interesting and rewarding activities:

1. Familiarize yourself with the definition of *regular tessellation* (p. 326). Encourage your child to find tessellations in your home, such as floor tile patterns, wallpaper patterns, and wall tile patterns. Have your child identify the shapes that make up the pattern.

2. Encourage your child to use the local library or the Internet to find examples of M. C. Escher's artwork.

3. If you have art software for your home computer, allow your child time to experiment with computer graphic tessellations. Encourage him or her to share the creations with the class.

Building Skills through Games

In Unit 10, your child will reinforce skills and concepts learned throughout the year by playing the following games:

Angle Tangle See *Student Reference Book,* page 306
Two players will need a protractor, straightedge, and blank paper to play *Angle Tangle.* Skills practiced include estimating angle measures as well as measuring angles.

Name That Number See *Student Reference Book,* page 329
This game provides your child with practice in writing number sentences using order of operations. Two or three players need 1 complete deck of number cards to play *Name That Number.*

241

As You Help Your Child with Homework

As your child brings assignments home, you may want to go over the instructions together, clarifying them as necessary. The answers listed below will guide you through some of the Unit 10 Study Links.

Study Link 10·1

1. rotation

2. translation

3. Answers vary.

4. Answers vary.

5. 114.534

6. 35.488

7. 0.0338

8. 31.7025

Study Link 10·2

1.

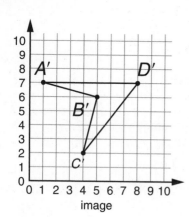

image

2.

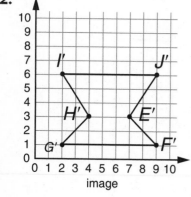

image

3.

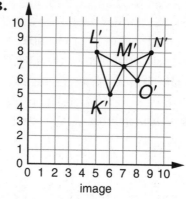

image

4. 0.8   **5.** 1.6   **6.** 8.9   **7.** 5.1

Study Link 10·3

1. 2   **2.** 1   **3.** 4

4. 6   **5.** 2   **6.** infinite

7. 2, 3, 5, 6, 9, 10   **8.** 2, 3, 6, 9

Study Link 10·5

Sample answers:

1. The paper clips are linked to one another.

2. The paper clips and the rubberband are linked.

3. All the paper clips are linked.

4. 60   **5.** 50   **6.** 63   **7.** 493

STUDY LINK 10·1 | **Tessellation Exploration**

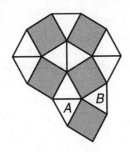

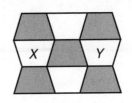

1. What transformation would move Figure *A* onto Figure *B*?

_____

2. What transformation would move Figure *X* onto Figure *Y*?

_____

3. Pick one or more polygons from the Geometry Template that you know will tessellate. In the space provided below, draw a tessellation made up of the polygon(s).

4. Tell whether the tessellation you drew is regular or semiregular. Explain how you know.

_____

_____

| **Practice** |

5. 5.67 * 20.2   6. 443.6 * 0.08   7. 6.76 * 0.005   8. 14.09 * 2.25

_____   _____   _____   _____

Translations

Plot and label the vertices of the image that would result from each translation.
One vertex of each image has already been plotted and labeled.

1.

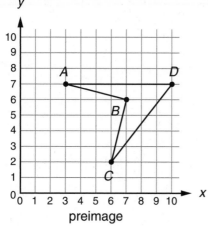

preimage

horizontal
translation

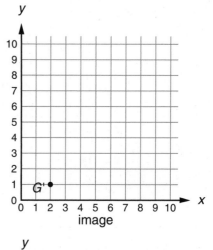

image

2.

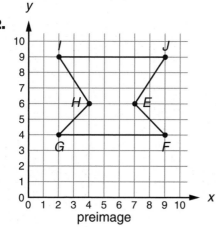

preimage

vertical
translation

image

3.

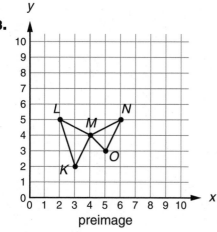

preimage

diagonal
translation

image

Practice

4. $\dfrac{25.6}{32}$ _____

5. $\dfrac{102.4}{64}$ _____

6. $\dfrac{41.83}{4.7}$ _____

7. $\dfrac{67.32}{13.2}$ _____

245

Rotation Symmetry

For each figure, draw the line(s) of reflection symmetry, if any.
Then determine the order of rotation symmetry for the figure.

1.

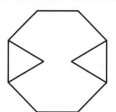

Order of rotation symmetry _____

2.

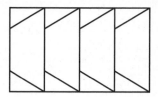

Order of rotation symmetry _____

3.

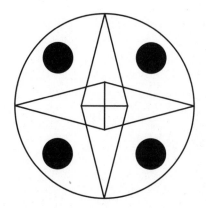

Order of rotation symmetry _____

4.

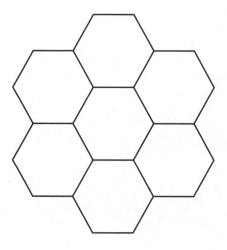

Order of rotation symmetry _____

5.

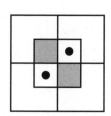

Order of rotation symmetry _____

6.

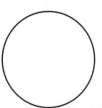

Order of rotation symmetry _____

Practice

Tell whether each number is divisible by 2, 3, 5, 6, 9, or 10.

7. 4,140 _____

8. 324 _____

247

A Topology Trick

SRB
184 185

Follow the procedure described below to tie a knot in a piece of string without letting go of the ends.

Step 1 Place a piece of string in front of you on a table or a desk.

Step 2 Fold your arms across your chest.

Step 3 With your arms still folded, grab the left end of the string with your right hand and the right end of the string with your left hand.

Step 4 Hold the ends of the string and unfold your arms. The string should now have a knot in it.

This trick works because of a principle in topology called **transference of curves.** Your arms had a knot in them before you picked up the string. When you unfolded your arms, you transferred the knot from your arms to the string.

Another Topology Trick

Follow the procedure described below to perform another topology trick that works because of transference of curves.

Step 1 Gather the following materials: 2 to 8 large paper clips, a strip of paper $1\frac{1}{2}$ by 11 inches, and a rubber band.

Step 2 Curve the strip of paper into an S-shape. Attach two paper clips as shown at the right.

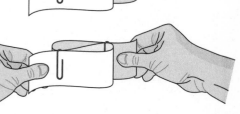

Step 3 Straighten the paper by holding the ends and pulling sharply.

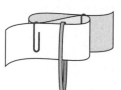

1. Describe your results.

_____

_____

2. Add a rubber band as shown. Straighten the paper.

Describe your results.

_____

_____

3. Try including a chain of paper clips as shown.

Describe your results.

_____

_____

Practice

Find the LCM of each pair of numbers by dividing the product of the numbers by their GCF.

4. 15 and 20   **5.** 10 and 50   **6.** 21 and 63   **7.** 17 and 29

_____   _____   _____   _____

251

STUDY LINK 10·6 | Family Letter

Congratulations!

By completing *Sixth Grade Everyday Mathematics,* your child has accomplished a great deal. Thank you for your support.

This Family Letter is intended as a resource for you to use throughout your child's vacation. It includes an extended list of Do-Anytime Activities, directions for games that you can play at home, a list of mathematics-related books to get from your library, and a preview of what your child might be learning in seventh grade.

Do-Anytime Activities

Mathematics means more when it is rooted in real-world situations. To help your child review many of the concepts learned in sixth grade, we suggest the following activities for you to do with your child over vacation. These activities will help your child build on the skills that he or she has learned this year and are good preparation for a seventh-grade mathematics course.

1. Practice quick recall of multiplication facts. Include extended facts, such as $70 * 8 = 560$ and $70 * 80 = 5,600$.

2. Practice calculating mentally with percents. Use a variety of contexts, such as sales tax, discounts, and sports performances.

3. Use measuring devices—rulers, metersticks, yardsticks, tape measures, thermometers, scales, and so on. Measure in both U.S. customary and metric units.

4. Estimate the answers to calculations, such as the bill at a restaurant or store, the distance to a particular place, the number of people at an event, and so on.

5. Play games like those in the *Student Reference Book.*

6. If you are planning to paint or carpet a room, consider having your child measure and calculate the area. Have him or her write the formula for area ($A = l * w$) and then show you the calculations. If the room is an irregular shape, divide it into separate rectangular regions and have your child find the area of each one.

7. Ask your child to halve, double, or triple the amount of ingredients needed in a particular recipe. Have your child explain how they calculated each amount.

8. Help your child distinguish between part-to-part and part-to-whole ratios in relation to the wins and losses of a favorite sports team. Ask him or her to decide which ratio is being used. For example, wins to losses (such as 5 to 15) or losses to wins (15 to 5) are part-to-part ratios. Part-to-whole ratios are used to compare wins to all games played (5 out of 20) or losses to all games played (15 out of 20).

9. Provide extra practice with the partial-quotients division algorithm by having him or her divide 3-digit numbers by 2-digit numbers, 4-digit numbers by 3-digit numbers, and so on. Ask your child to explain the steps of the algorithm to you as she or he works through them.

253

Building Skills through Games

The following section lists directions for games that can be played at home. Regular playing cards can be substituted for the number cards used in some games. Other cards can be made from 3" by 5" index cards.

Name That Number See *Student Reference Book* page 329.
This game provides practice in using order of operations to write number sentences. Two or three players need a complete deck of number cards.

Fraction Action, Fraction Friction See *Student Reference Book* page 317.
Two or three players gather fraction cards that have a sum as close as possible to 2, without going over. Students can make a set of 16 cards by copying fractions onto index cards.

Name That Number

Materials　　☐　4 each of number cards 0–10 and
　　　　　　　 ☐　1 each of number cards 11–20

Players　　 2 or 3

Skill　　　　 Naming numbers with expressions

Object of the game　To collect the most cards

Directions

1. Shuffle the deck and deal five cards to each player. Place the remaining cards number-side down on the table between the players. Turn over the top card and place it beside the deck. This is the **target number** for the round.

2. Players try to match the target number by adding, subtracting, multiplying, or dividing the numbers on as many of their cards as possible. A card may only be used once.

3. Players write their solutions on a sheet of paper. When players have written their best solutions:

   ◆ Each player sets aside the cards they used to match the target number.

   ◆ Each player replaces the cards they set aside by drawing new cards from the top of the deck.

   ◆ The old target number is placed on the bottom of the deck.

   ◆ A new target number is turned over, and another round is played.

4. Play continues until there are not enough cards left to replace all the players' cards. The player who has set aside the most cards wins the game.

Fraction Action, Fraction Friction

Materials ☐ One set of 16 *Fraction Action, Fraction Friction* cards. The card set includes a card for each of the following fractions (for several fractions there are 2 cards): $\frac{1}{2}, \frac{1}{3}, \frac{2}{3}, \frac{1}{4}, \frac{3}{4}, \frac{1}{6}, \frac{1}{6}, \frac{5}{6}, \frac{1}{12}, \frac{1}{12}, \frac{5}{12}, \frac{5}{12}, \frac{7}{12}, \frac{7}{12}, \frac{11}{12}, \frac{11}{12}$.

☐ One or more calculators

Players 2 or 3

Skill Estimating sums of fractions

Object of the game To collect a set of fraction cards with a sum as close as possible to 2 without going over 2.

Directions

1. Shuffle the deck. Place the pile facedown between the players.

2. Players take turns.

♦ On each player's first turn, he or she takes a card from the top of the pile and places it number-side up on the table.

♦ On each of the player's following turns, he or she announces one of the following:

Action This means the player wants an additional card. The player believes that the sum of the fraction cards he or she already has is *not* close enough to 2 to win the hand. The player thinks that another card will bring the sum of the fractions closer to 2, without going over 2.

Friction This means the player does not want an additional card. The player believes that the sum of the fraction cards he or she already has *is* close enough to 2 to win the hand. The player thinks that there is a good chance that taking another card will make the sum of the fractions greater than 2.

Once a player says *Friction*, he or she cannot say *Action* on any turn after that.

3. Play continues until all players have announced *Friction* or have a set of cards whose sum is greater than 2. The player whose sum is closest to 2 without going over 2 is the winner of that round. Players may check each other's sums on their calculators.

4. Reshuffle the cards and begin again. The winner of the game is the first player to win five rounds.

Vacation Reading with a Mathematical Twist

Books can contribute to learning by presenting mathematics in a combination of real-world and imaginary contexts. Teachers who use *Everyday Mathematics* in their classrooms recommend the titles listed below. Look for these titles at your local library or bookstore.

For problem-solving practice:

Math for Smarty Pants by Marilyn Burns, Little, Brown and Company, 1982.
Brain Busters! Mind-Stretching Puzzles in Math and Logic by Barry R. Clarke, Dover Publications, 2003.
Wacky Word Problems: Games and Activities That Make Math Easy and Fun by Lynette Long, John Wiley & Sons, Inc., 2005.
My Best Mathematical and Logic Puzzles by Martin Gardner, Dover Publications, 1994.
Math Logic Puzzles by Kurt Smith, Sterling Publishing Co., Inc., 1996.

For skill maintenance:

Delightful Decimals and Perfect Percents: Games and Activities That Make Math Easy and Fun by Lynette Long, John Wiley & Sons, Inc., 2003.
Dazzling Division: Games and Activities That Make Math Easy and Fun by Lynette Long, John Wiley & Sons, Inc., 2000.

For fun and recreation:

Mathamusements by Raymond Blum, Sterling Publishing Co., Inc., 1997.
Mathemagic by Raymond Blum, Sterling Publishing Co., Inc., 1992.
Kids' Book of Secret Codes, Signals, and Ciphers by E. A. Grant, Running Press, 1989.
The Seasons Sewn: A Year in Patchwork by Ann Whitford Paul, Browndeer Press, 1996.

Looking Ahead: Seventh Grade

Next year, your child will:

◆ increase skills with percents, decimals, and fractions.

◆ compute with fractions, decimals, and positive and negative numbers.

◆ continue to write algebraic expressions for simple situations.

◆ solve equations.

◆ use formulas to solve problems.

Thank you for your support this year. Have fun continuing your child's mathematical experiences throughout the summer!

Best wishes for an enjoyable vacation.